Machine Learning for Engineers

Marcus J. Neuer

Machine Learning for Engineers

Introduction to Physics-Informed,
Explainable Learning Methods for
AI in Engineering Applications

 Springer

Marcus J. Neuer
Research & Development
innoRIID
Düsseldorf, Germany

ISBN 978-3-662-69994-2   ISBN 978-3-662-69995-9 (eBook)
https://doi.org/10.1007/978-3-662-69995-9

I dedicate this book to my great love, Paula!

Preface

Dear reader! This book originated from the lecture series „Data Mining in the Context of Technical Processes", which I was able to give at RWTH Aachen in recent years. Often, there was great interest from the students with regard to the learning methods. It became clear to me, even though there is no shortage of good computer science literature on this, that it makes sense to describe the application, design, and operation of such methods from the perspective of technical applications. The practical implementation plays a particularly important role in this.

The book is intended to provide a starting point for students who have had little contact with machine learning so far, and to help them with a selected collection of runnable code fragments to work on their own problem situations with learning methods. In doing so, it should convey enough mathematical basics to create confidence in the methodology, as well as present the programming sufficiently completely to easily build upon it. The theory shown here therefore only represents an introduction.

From the first chapter onwards, the focus is on physics-informed methods and the aspect of algorithmic explainability. Both the structure of the book and the contents of the individual chapters have been aligned with this thematic objective. Thus, the presentation of the mathematical foundations includes a focus on stochastic processes, in order to later integrate these into the physics-informed learning methods. This requires a consistent inclusion of uncertainties, analytical expressions, and semantic tools.

For reasons of better readability, we predominantly use the generic masculine in this book. This always implies both forms, thus including the feminine form.

The book is divided into three larger sections. Chapters 1 to 3 form a kind of introductory. They deal with the handling of data, mathematical tools to describe them, and ultimately methods to adapt them purposefully.

Chapters 4, 5 and 6 deal with Machine Learning, starting with supervised learning methods, moving on to unsupervised methods, and then to the idea of physics-informed learning. Each method is not only presented, but also supported with code examples. This application at the programming level is of great importance, as it deepens understanding and enables future applicability in the first place.

There are approaches that we have deliberately left out. Support Vector Machines (SVM), Kohonen's Self-Organizing Map (SOM), or even Restricted Boltzmann Machines (RBM) would have certainly fit into the context. However, a sufficiently detailed presentation would have exceeded the scope of this introduction and should therefore be revisited elsewhere in the future.

In all chapters, explainability and basic physical understanding are repeatedly addressed as themes. Chapter 7 concludes the book and addresses this aspect from various perspectives. We not only show helpful semantic tools to store context and knowledge references, but we also devote ourselves to the question of which data technologies and strategies support explainability.

Düsseldorf   Marcus J. Neuer
July 2023

Acknowledgments

Over time, various students have supported me in the realization of the book by proofreading, following the examples, and debugging the source code. I would like to express my sincere thanks for this, even though I cannot list everyone by name here.

Just as I would like to thank the Springer Nature team for their support, advice, and patience in the creation of this book.

Andreas Quick, Thomas George, and Tobias Seitz from iba AG often discussed with me about the practical aspects of learning methods and the requirements of industrial customers. I am very grateful for these stimulating discussions and insights to this day.

Peter and Christian Henke have given me the opportunity at innoRIID GmbH to specialize many of my algorithmic approaches for real products and thus commercialize them. In doing so, I was able to gain valuable insights into which methods are truly robust and which are less suitable for practical use. For this, I thank them and also the entire team at innoRIID.

I would also like to express my heartfelt thanks to the many colleagues at the Operational Research Institute (BFI) in Düsseldorf. Over the past few years, I have found a scientific home here that has allowed me to get to know and also learn to teach many new topics. Dr. Alexander Ebel certainly infected me with his passion for semantic technologies. I am grateful to him for the joint projects in which we were able to bring agent technologies and semantic concepts into real use in the industry. Norbert Link often stood by my side to discuss the meaningfulness of an algorithm. From him, I have received many insights to critically question methods and detect errors. Norbert Holzknecht aroused in me a fascination for industrial data systems and encouraged me to also use unconventional solutions. Dr. Andreas Wolff was always there to advise and assist me, especially when it came to complicated mathematical concepts. I am grateful to him for the many discussions and brainstorming sessions that have expanded my horizon each time. Finally, I would like to mention Prof. Dr. Harald Peters, without whom I probably would never have resumed my academic interests and who has always supported me to this day.

A final big thank you goes to the Institute for Theoretical Physics I of Heinrich-Heine-University Düsseldorf. Specifically, Prof. Dr. K.-H. Spatschek and Dr. E. Laedke have taught me fundamental algorithmic and mathematical tools that accompany me in my profession to this day, which makes me deeply grateful.

I would like to thank my parents, who made so many things possible for me in life and are always there for me. They supported every one of my life dreams and thus also have a large share in this work.

This book is not only dedicated to my wife Stephanie Paula Neuer, but I also thank her for her loving support, for being the wind in my sails, and of course for the many hours she spent helping me find and eliminate errors. Without her, this book would not have been possible.

in Düsseldorf   Marcus J. Neuer
July 2023

Contents

List of Abbreviations

AE   Autoencoder
DGL   Differential equation
IG   Information Gain`
KNN   Artificial Neural Network
LSTM   Long Short-Term Memory
MDN   Mixture-Density-Network
MLP   Multi-Layer Perceptron
NN   Neural Network
PCA   Principal Component Analysis
PINN   Physically-informed neural network
Xtest   Test data of the input variables
Xtrain   Training data of the input variables
Ytest   Test data of the label/target variables
Ytrain   Training data of the label/target variables

Chapter 1
Data as the Basis of Models

Keywords Data Mining · CRISP-DM · Scales

Machine learning has achieved success in many applications in recent years. The possibilities it opens up for us are numerous. Novel photo filters calculate our face to look younger or ensure that we always look into the camera lens. Knowledge is accessed faster than before through intelligent chatbots. In this sense, data-based models and learning methods are currently changing our environment sustainably. Many people react with skepticism. To some extent, this is justified, because artificial intelligence (AI), which is often mistakenly equated with its subfield of machine learning, cannot solve every problem. The expectation of AI is exaggerated and utopian (Fig. 1.1).

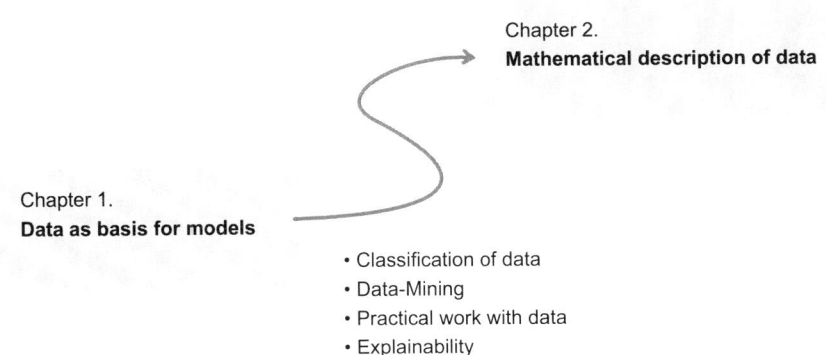

Fig. 1.1 Chapter overview

© The Author(s), under exclusive license to Springer-Verlag GmbH, DE, part of Springer Nature 2025
M. J. Neuer, *Machine Learning for Engineers*,
https://doi.org/10.1007/978-3-662-69995-9_1

Data is the fuel for learning methods. It is important to understand how data can be characterized. This chapter therefore shows how we can describe and work with them in a structured way. We introduce Data Mining and the Cross Industry Standard Process for Data Mining (CRISP-DM) as systematic processes, with a clear focus on the practical aspects of industrial implementation. To understand the program codes in the book, we briefly introduce the programming language Python and explain its most important libraries.

Machine learning for engineering sciences focuses on the context of technical processes and production systems. It therefore primarily has a practical reference. We systematically present the topic in the context of this book and place great value on clear examples and comprehensible practical applications. Some of these applications and examples have been simplified. We present the theory to motivate confidence in a method or to highlight a specific mode of action. However, the mathematical foundations, as shown here, do not claim to be formulated in a closed manner. Rather, they should help to make the underlying mechanisms comprehensible on a mathematical basis.

1.1 Data-Based Modeling

1.1.1 The Concept of Model

Why is machine learning interesting at all? For this, we need to go a bit more into details of models. In the natural and engineering sciences, there has been a deep appreciation for models for many years. They imitate reality at a lower level of detail and help us understand complex relationships. Processes become predictable through models.

First-Principle Models
Many models originate from axioms and the laws of nature. They link physical quantities with each other and thus allow a direct understanding of the relationships. Because of this property, they are called First-Principle models. They are captured by compact mathematical equations. Differential equations, conservation laws up to state models of control technology are examples of this.

A well-known first-principle model illustrates this a bit closer: the law of gravity. The force that two masses m_1 and m_2 exert on each other is proportional to the product of these masses and inversely proportional to the square of their distance r,

$$F \propto \frac{m_1 m_2}{r^2}. \tag{1.1}$$

This model allows us to directly understand the relationship between the masses and their distance. It is also compactly formulated. Eq. (1.1) states what will happen if we, for example, double the mass m_2. On the other hand, if we find that the force has become four times smaller than before, we can say why by measuring r and knowing m_1 and m_2: because the radius has doubled.

Let's highlight the important features of models again: a) They help us understand complex relationships as they link variables together and b) they predict the behavior of systems.

Data-based Models
Since machines are based on scientific principles, models help us identify problems in industrial production. Many technical processes are combinations of several processes. Often, the individual processes are so complex that a complete representation is only possible to a limited extent, even with reduced first-principle models. Especially the original bottom-up approach, deriving relationships from axioms or basic laws, is difficult.

This is where data-based modeling comes into play. Data exists from many technical processes. What happens in the processes can be recorded at least within the accuracy of the sensors. Assuming all relevant data is captured and we have both the influencing variables and the size we want to understand or predict, then the actual dependency is contained in the data and can be extracted from it. This is the basic assumption of data-based modeling.

In contrast to deriving a model with the bottom-up approach, data-based modeling begins with the observation of the processes. This is top-down, as the process serves as the starting point and the details are only discovered afterwards. Subsequently, statistical tools are used to set up simple models. As complexity increases, machine learning methods are used. They form a separate subgroup of data-based modeling. Yet, there are difficulties and open questions:

- **Choice of variables.** Is the dynamics we want to capture even captured by the data?
- **Quality of measurement.** Are the sensor data accurate enough to represent the problem?
- **Amount of data.** Do we have enough measurement points available to set up the desired model?

1.1.2 White-Box, Grey-Box, and Black-Box Models

The Characterization in First-Principle and data-based models is oriented towards the model's origin. Another property is the traceability of a model. Here, the following categories are distinguished:

- **White-Box Models.** Any model that is completely traceable and explainable is called a White-Box model. We can look into the model and understand how it works. First-Principle models are White-Box models. Data-based approaches such as linear regression can also be White-Box models.
- **Black-Box Models.** If traceability is not possible and thus the internal mechanisms are not known, then it is called a Black-Box model. Such models can predict processes, but they do not allow any statement about the relationship

between input and output variables. Consequently, it is difficult to trust Black-Box models. Machine learning algorithms are often accused of belonging to this category. However, this is not necessarily true. There are methods, as we will learn later, that help us to investigate models for their internal mechanisms and thus move from the Black-Box character to a real understanding and trust.

- **Grey-Box Models.** A third variant is Grey-Box models, which are partially traceable. They use input variables whose influence is known, and additional variables whose effect cannot be captured. Monte Carlo methods are counted in this category because they contain an analytical core, e.g., a differential equation, and simulate this with stochastic variables. Since the latter are random processes, at least part of the Monte Carlo simulation is unpredictable.

1.1.3   Criticism of Data-Based Modeling

The lack of reference to natural laws is a disadvantage of data-based modeling. Some practitioners and users criticize data-based models because of their difficult traceability. They base their criticism on a purely Black-Box character. The acceptance of machine learning was initially affected by this. However, many of these critics are based on false assumptions.

The diversity of models has increased over the years. Machine learning algorithms can be made understandable and explainable. They now belong to the area of Grey-Box models. The integration of physical First-Principle models as part of data-based models is unknown to many users. However, these approaches have many successes in the industry to show.

Learning methods can be sampled. By deliberately disturbing the input variables, one can check how the algorithm behaves. Thus, we can identify analytical relationships for non-deterministic neural networks. For most learning methods, a basis for explainability can be achieved with the help of tools from stochastic.

1.2   Classification of Data

1.2.1   Historical Classification of the Term Data

Handwritten lists, analog photos, post-its, or books are referred to as data—a term that is neither new nor a modern invention. It goes back several hundred years and is closely linked to the concept of number and the concept of measuring and comparing.

People began early to introduce comparative measures to keep track of agricultural activities. They could determine the stock of a cornfield and estimate the possible harvest yields, which was important for food production. By comparing

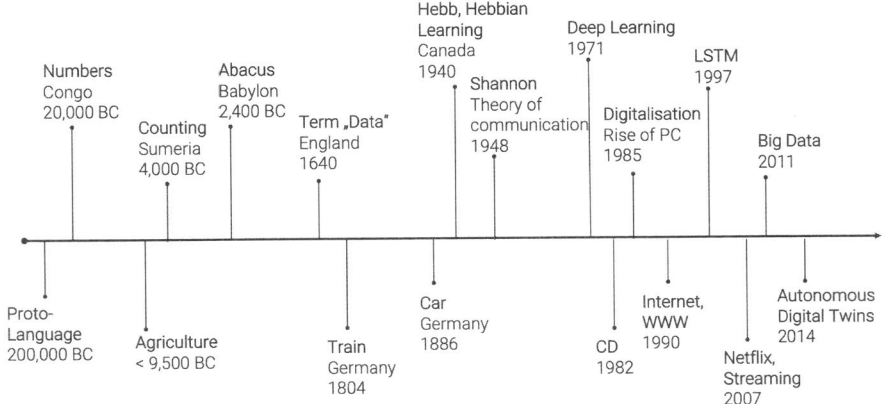

Fig. 1.2 Timeline with some important events

with the written information from previous years, it could be determined whether the yield was declining or not.

More generally, data is collected with the aim of monitoring states or processes and ultimately influencing them. In Fig. 1.2, some important events are shown to give an impression of how data has also influenced important milestones in industrial development. Please note, however, that this timeline is not complete.

1.2.2   Data in Our Present Time

Data is everywhere today. It is generated intentionally or unknowingly, planned or accidentally. Often, a lot of data is created simply because it was not considered in advance which data is really needed to be collected—an approach that is not recommended, but often unavoidable. We often simply do not know which information will be important later on. That is the reason why many data is actually never used.

We want to systematically approach the topic of data evaluation. For this, we need a basic understanding of how to describe, characterize, and structure data.

In today's time, data is ubiquitous because it is the backbone of many everyday technologies. How different data can be and that there are various types of them seems evident just by looking at text messages, websites, smartphones, or medical records. Consider what data is relevant in the following processes:

- Smartphones, which guarantee ubiquitous access to the World Wide Web,
- Social media like Facebook, which represent data collections of their users,
- Services for sending messages.

For these examples, heterogeneous combinations of different data types are always needed. In the next sections, we therefore explain how to describe and arrange data.

1.2.3   Structural Classification

A first, rough division of data can be made based on their structuring.

Structured Data
Each of us has used a table before. They have column names and a defined row structure. Their size is usually fixed and the type of each field is precisely defined. This type of data is referred to as structured.

> **Example: Data from a group of people**
> Imagine you have a group of 30 people. You ask them to identify themselves. To do this, you ask each person to enter certain information about themselves into a questionnaire with fixed input fields. The fields ask for things like name, date of birth, and address. Anyone who fills out the form is automatically forced to adhere to this structure. ◄

Structured data is therefore characterized by the presence of such a predefined schema, which is also referred to as a **data model**.
Further examples of structured data are:

- Data from technical processes,
- Bank transactions,
- Data for search engine optimization.

Unstructured Data
If there is no data mode, then we speak of unstructured data. These are defined as information that cannot be mapped by a fixed schema. So, you don't know for sure in what form the data will be presented. Staying with the above example, this corresponds to the case where you ask all 30 people in the room to give you some personal information. Some will start writing something, others will give you a copy of their ID card, and others will simply contribute a few personal photos. The data is unstructured.

- Photographs,
- Books,
- Health records,
- Poetry albums

are examples of unstructured data.

Semi-structured Data
Now there is another, often encountered type of data, which can be seen as a mixture of both of the above extreme cases: semi-structured data. Here,

some predefined basic form exists, so to speak a metamodel, according to which one expects the data to be structured, but in some subdivisions, one deliberately allows completely unstructured data (Fig. 1.3).

In our above example, the group of people would be asked to fill out a form with predefined fields and fields to be edited freely. So they would give their names and addresses, but also have some space to enter any additional information. Some might list their hobbies, others add three pictures, and others draw something. Part of the data is thus structured, another part unstructured. This type of data is then referred to as semi-structured.

Some examples of this are:

- Emails,
- Websites,
- Tweets,
- the memory of Digital Twins.

The structure of our data determines the choice of a suitable database architecture. Thus, SQL databases can process structured data particularly well; however, they have difficulties with unstructured or semi-structured data. In contrast, there are various Not-Only-SQL-(NoSQL-)database concepts, that were specifically developed for the storage and management of semi-structured data.

1.2.4   Quantitative Categorization of Data

Data can further be divided into two more categories: continuous and discrete data.

Continuous Data
Data that can assume any value within a certain range are referred to as continuous data. Measuring a length with a ruler is an example of (quasi-)continuity. Any arbitrary number (0.1, 0.01, 0.001, …) is theoretically possible. Sometimes continuous data is synonymously seen with the term **analog data**.

Discrete Data
Discrete data refers to individual, clearly separable data points that can be counted and are often represented by whole numbers (0, 1, 2, 3, 4, 5,…). An example of discrete data is the number of people on a train. Since there is no such thing as half a person, this can only be described with discrete numbers. **Digital data is discrete.**

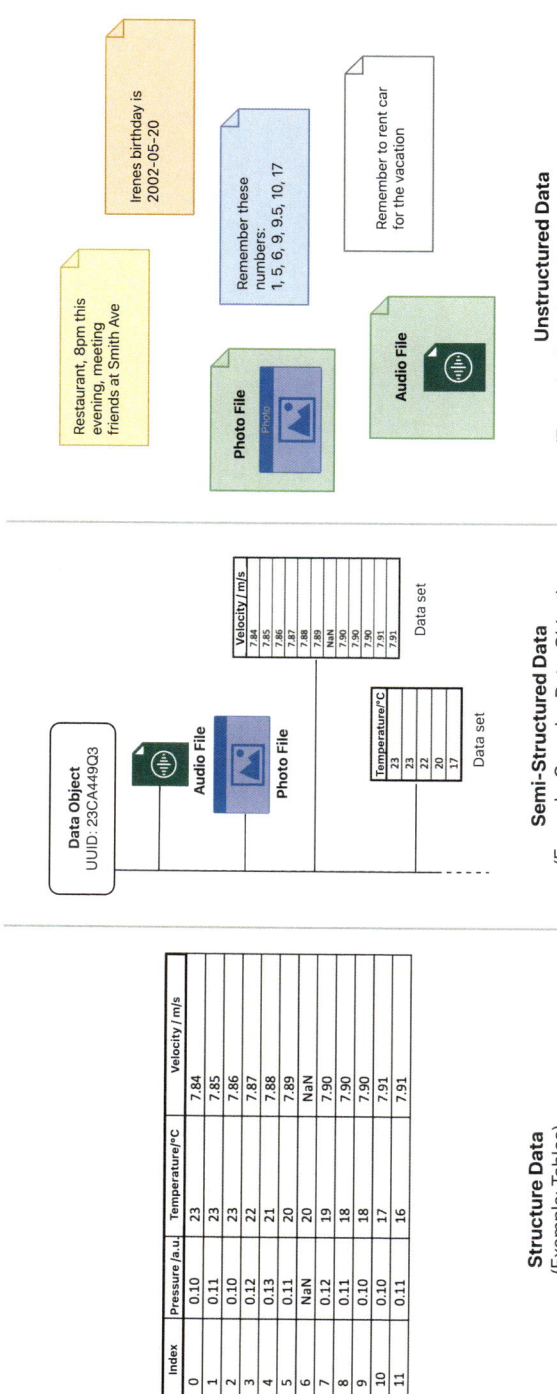

Fig. 1.3 From left to right: Structured, semi-structured, and unstructured data

1.2.5   Qualitative Categorization of Data

We can add another level of description to better understand data. This considers the nature of the data itself: Are they letters or numbers, months or lengths? The following qualitative types can be distinguished:

Nominal or attributive data
Descriptive data is called nominal—they carry a name. Terms in lists, such as colors, street names, and descriptions, represent cases of nominal data. In programming languages, the most common type used for nominal data is the **string.** They are also difficult to measure. They often require manual input. If we consider the example with the 30 people and the forms, the entry fields first name, last name, and address are of course of a nominal nature.

Ordinal Data
If we can put the data in a logical order, they are ordered and we call this type ordinal data. Obvious ordinal data are e.g. the names of the months ("January", "February", …). Please note that ordinal data also have a nominal component, as the example with the month names shows. The ability to sort is thus an additional capability of this form of data.

Cardinal Data
We can add and subtract cardinal data, multiply and divide. It is possible to apply arithmetic rules to them. Number spaces are cardinal data. With month names (which are ordinal and not cardinal), you cannot perform addition, for example. That's why we use the separate term cardinal, (lat. *numeri cardinale,* special numbers). In programming languages, the data type **integer** represents the natural numbers, and the data type **float** (floating point) captures the real numbers (floating point numbers).

Binary Data
Binary decisions are "Yes"/"No", 0/1 or "True"/"False". The specific variant with "True"/"False" is also referred to as a **boolean** decision. The data type in programming languages is called **bool** or **boolean.**

Labels and Target Sizes
A very important, more general type of qualitative data are the so-called **labels.** These are a specific set of nominal, ordinal, cardinal, or binary information used in supervised machine learning as training targets for the algorithm. The term label will accompany us in further chapters; therefore, it is highlighted here. ◄

1.2.6 Time Series

A data series, in which the individual data points are indexed and ordered by time, is referred to as a time series. Time series can be categorized as cardinal, discrete data. This type of data is primarily found in technical processes. In physics, most dynamic equations (the model) are formulated as derivatives with respect to time. All these systems can be monitored by measuring time series, e.g., to verify the validity of the model. If you consider a piece of steel that cools down after heating and rolling, and you take measurements every 5 min, you get a time series.

In the analysis of technical processes, the majority of cases involve combinations of time series and individual nominal attributes. They are at the heart of technical data mining (Fig. 1.4).

Time series can also be evaluated in the frequency domain, which is well suited for repetitive and oscillating processes. The scope of time series analysis becomes clear when looking at the book by J. D. Hamilton [5], in which many different model approaches to time series are discussed.

1.2.7 Scales

The consideration of the above qualitative and quantitative categorization of data is related to the concept of scale [4]. A scale divides a set of values or properties into measurable sections. It is a benchmark for our data. Depending on their qualitative characterization, we distinguish, for example, nominal scales or ordinal scales. The scale depends on its data type and the ability to divide the data collection into measurable sections. The difference between nominal and ordinal scales becomes clearer with the following examples:

Fig. 1.4 A simple scheme to describe data

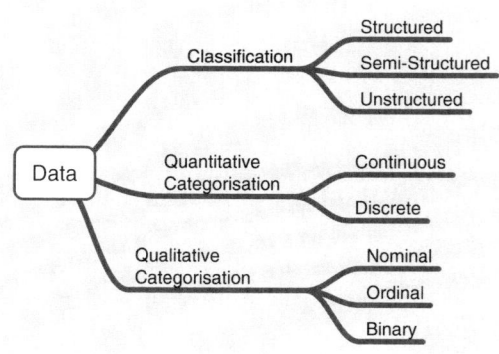

Example: Teacher and Students
Imagine you are a teacher and teach a class with N students. Each student has a first name. When you write down the first names, you record nominal data. What can you measure from these data? In fact, you can determine the frequency with which a name occurs. Ultimately, you know that there are four people named Michael and two named Julia, while the remaining names occur exactly once. ◄

Example: Students' Birthdays as Temporal Ordering Variable
If you go one step further and also ask for the students' birthdays, you extend your nominal data collection by an ordinal variable, namely a date. Now you can sort the individual people by their birthday. You can even distinguish people with the same name. ◄

Nominal data can therefore only be compared with each other or with a reference. For variables of this category, you only have the operations $=$ or $\neq$ available. The ordinal data, on the other hand, also have the operations $<$ and $>$.

Consequently, for cardinal data, there is another scale, the cardinal scale. On it, mathematical operations $+$, $-$, $*$ and $/$ are executable. Since this allows the comparison of different intervals, this scale is also called interval scale. If a defined zero point is additionally introduced, we can not only define intervals, but also calculate ratios. A cardinal scale with zero point is therefore called ratio scale. Tab. 1.1 summarizes the different scales and their properties once again compactly.

Example: Analog and digital temperature measurement
When a weather station measures temperatures, it retrieves the temperature value from the sensor at fixed intervals. These are discrete, cardinal

Tab. 1.1 Overview of scale levels

Scale level	Math. Operation	Measurement sizes	Stat. Sizes	Example
Nominal scale	$=, \neq$	Frequency, Sorting	Mode	Names
Ordinal scale	$=, \neq, <, >$	Frequency Sorting	Median	Months
Cardinal scale – Interval scale	$=, \neq, <, >,$ $+, -, *, /$	Frequency, Sorting, Distance	Arith. Average	Temperature/°C
Cardinal scale – Ratio scale	$=, \neq, <, >,$ $+, -, *, /, 0$	Frequency, Sorting Distance, Zero point	Geo. Average	Temperature/ K

data. With a liquid thermometer, this recording is on the cardinal scale, but analog. ◄

Example: Seismograph
The recording of vibrations of the earth's surface is done with a seismograph. It consists of a spring with an inert mass. As soon as the ground is shaken, a recorder records the relative movement of the ground to the mass. This is an example of continuous measurement. The data from the seismograph are continuous time series on a cardinal scale. ◄

An important task of machine learning in practice is the interplay of data on their various scales. We use our definitions from the previous sections to consider some more examples and classify them:

Example: Lottery
A weekly drawing of lottery numbers is a sample from an N-element set. The data generated in this way are discrete. It is an upwardly bounded nominal scale of defined numbers. Summation and subtraction are not meaningfully applicable here. Also, the order of the numbers to be drawn does not matter—but the order in which they are drawn does. ◄

Example: Books as a collection of data
The text in books is a discrete sequence of nominal information—captured by separate letters that form words and separate words that form sentences. ◄

Note the scales you are working on. Scales also have an influence on your interpretation of data: A change of 5 °C from 20 °C to 25 °C, corresponds in the Celsius scale (an interval scale) to a change of 25 %. In the Kelvin scale (a ratio scale with an absolute zero point) the same change from 293 K to 298 K is only a difference of 1.7 %. It may be that your own perception, however, corresponds more to the Celsius scale—because this scaling is simply better oriented to our reality of life.

This last point will be discussed again in the next chapter when we consider the normalization of data. Then we artificially intervene in the scaling of our data to optimize the scale for a problem.

1.3   Data Mining as a Systematic Process

1.3.1   What is Data Mining?

The term data mining is based on the idea of mining rock. If you think of data as a difficult, opaque material, this analogy makes sense. The corresponding methods for extracting information from this data rock are then imagined as mining processes.

The analogy to mining is indeed appealing, but it harbors an interesting discrepancy to the practice of data analysis: It assumes that the data already exists and only needs to be evaluated. In fact, however, the task of a data miner also includes defining suitable ways to obtain the data. Data mining is understood as the analysis of a data set with the following objectives:

- **Detection.** Detection means perceiving something specific, respectively noticing it. In the context of data, it is often necessary to find outliers, anomalies or other relevant and unexpected behaviors that are responsible for problems in the technical process. It is often a challenge to clearly define what the normal (desired) behavior of a system is.

Example: Anomaly in Aircraft Control
Consider the control of an airplane. It has actuators such as thrust, elevators and rudders, tailplane and flaps. A pilot can influence the movement of the aircraft with these parameters. To measure the state of the aircraft and the result of possible control inputs, sensors are also present. They measure, among other things, the pressure and can derive variables such as speed or altitude from it. If a control intervention is made, the individual sensors can react differently.
If you push the control stick forward, the aircraft tilts and the pressure begins to rise. A situation that is normal. However, if you notice a decrease in pressure with the same control input, you have an anomaly. Possibly the associated sensor is defective. ◄

- **Classification.** An important goal of data mining is to group data elements into classes. This is especially important in industrial problems. Here it happens that products have to meet quality classes or a product has to meet the acceptance criteria of a customer.

Example: Classification of Good and Bad Products
In the production of screw heads, an employee notices that the number of faulty threads has increased over the last few days. Therefore, he divides the

products into two groups, good and bad threads. He decides this based on the condition of the burr on the thread, which is a subjective criterion. ◄

The example shows the greatest difficulty at this point: the meaningful and complete definition of when something belongs to a category.

- **Agglomeration.** Uncovering hidden structures and clusters in the data is another goal of data mining. They help to develop a deeper understanding of the underlying dynamics. This often leads to the optimization of processes.

Example: Agglomeration of Good and Bad Products
We extend the last example and assume that all bad screws were produced at similar pressure values and similar temperatures. They therefore accumulate in the data space to form a so-called cluster. If we know these clusters, we can predict for future products whether the production will be successful or not. ◄

- **Association.** If dependencies are recognized in the data, rules and predictions can be derived from them. This is then referred to as association.

Example: Association of Good and Bad Products
Again, we extend the previous example. We know the parameters pressure and temperature that lead to bad products. Now we set up rules for good products from this by defining limit values for the critical variables—we therefore associate the terms good and bad with corresponding permissible data ranges. Whenever the screws come too close to these limit values or exceed them during their production, a more detailed examination is necessary. ◄

- **Regression.** As soon as the course of a variable can be predicted from the other variables, we speak of regression. It represents nothing more than a model of the process described by the variable.

Example: Exercise Bike
In the gym, there are stationary bikes. Although the bikes themselves do not move, a speed and a distance covered are determined from the rotation of the pedals. For this purpose, a real distance covered is simulated from the sensor data via regression models. ◄

1.3.2 Steps to Perform Data Mining

This objective is achieved through a specific (linear) workflow, which represents the actual definition of the tasks associated with data mining:

- **1. Focusing.** Here, the data required for the analysis are selected and collected. They are grouped and, if necessary, enriched with knowledge.
- **2. Cleaning.** Data can be distorted, contain false values or outliers. As a critical element of preprocessing, cleaning specifically also refers to techniques for dealing with incomplete data.
- **3. Transformation.** Often, data must be transformed in one way or another, with some transformation steps being necessary:
  - Rescaling, rebinning or resampling,
  - Normalization,
  - Differentiation (1st derivative, 2nd derivative, ...),
  - Integration and moving average,
  - Fourier transform (or Mellin, Laplace, Lagrange transform),
  - Wavelet transformation,
  - Determination of probability densities and histograms.

- **4. Analysis.** The most exciting part of a data mining activity is certainly the analysis itself. Here, appropriate algorithms are applied to evaluate the data. These algorithms will be explained in later chapters and include methods of analytical interpretation, supervised and unsupervised learning.
- **5. Evaluation and Reflection.** Based on the analysis result, it is necessary to critically question the result and make an assessment of its validity compared to the actual technical process.

Steps 1–3 are commonly also referred to as **Preprocessing**.

1.3.3 Drawing Conclusions from Data

The final step, the evaluation of the results, is necessary to critically reflect on the findings. In Fig. 1.5, the consumption of ice cream and the number of serial offenses are shown, normalized to a common maximum. Note the high correlation of these data, i.e., a great similarity in the shape and dependence of both variables over the course of the year. Can we already derive meaningful results from these two curves? Let's try out some causal dependencies and formulate what we could conclude from the mere observation of the data:

1. Ice cream consumption leads to violent crimes.
2. Criminals eat a lot of ice cream.

Both statements are of course misleading. Even if the data suggest otherwise, there is no causal relationship between the two processes. The misinterpretation arises

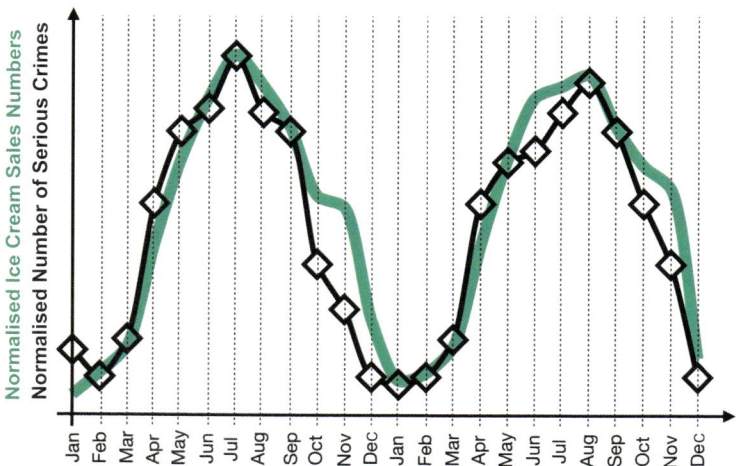

Fig. 1.5 Two different (normalized) time series: sales of ice cream (green) and number of violent crimes (blue) in a city

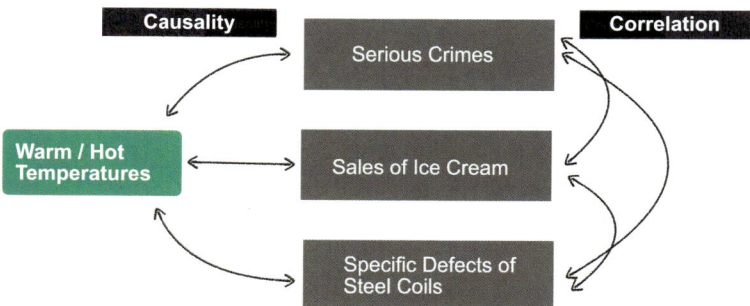

Fig. 1.6 A single cause can generate several similar data series without these being causally related to each other

from the difference between correlation and strong causality. One cannot simply derive a causal dependency from the observation of a correlation in the data. However, there could be a hidden cause for both curves—the causal source—that leads to the visible correlation. In this case, the cause is the temperature.

Fig. 1.6 illustrates the true dependency. High temperatures lead to high crime rates because they lower the tolerance for aggression. Of course, high temperatures also cause ice cream sales to rise because more and more people crave refreshment. In other technical contexts, high temperatures can lead to a lower number of certain defects.

This example is an association task. Starting from the identification of the root cause, we derive a rule for predicting a certain variable. Thus, it is possible to predict the consumption of ice cream or the crime rate by closely observing the temperature.

1.3.4   Industrial Data Mining and the CRISP-DM

The cross-industry standard process for data mining (Cross-Industry Standard Process for Data Mining, CRISP-DM), as presented in [10] among others, is a unified approach to the application of data mining in industrial projects. CRISP-DM is a process model. Such models help to establish a certain methodology in a company systematically.

- **Phase 1. Understanding the business model.** Companies need a special incentive to develop new methods and techniques. They must be profitable. Therefore, it is of great interest to determine the economic benefit and the business model behind the data mining task. Where does the mining activity add value, e.g., through money or energy savings or by avoiding errors or downtime? S. Aggarwal and N. Manuel assess in [2] the requirements from the business model as crucial for successfully implementing data-based solutions in the long term.

  It is also important for business understanding to quantify the Return on Investment (ROI). The decisive element here is the time it takes for the initial investment to pay off. Imagine you are spending resources and material costs to advance a data mining project. If this investment only pays off after twenty years or yields a negligible profit, it may not be worth pursuing at all.

  The result of Phase 1 must be a set of important **performance indicators** (Key Performance Indicators, KPI) (KPI) that measure the impact of the applied method. The KPIs are used throughout CRISP-DM for monitoring and improving processes.

- **Phase 2. Understanding the data.** Equally important is understanding the data needed to perform a particular data mining task. In the operation of technology companies or production facilities, this is often associated with answering the following questions:
  - What data do we need for the task?
  - What do these data describe? What processes are behind these data?
  - What do the data look like? Here you should manually visualize and check the data.
  - Are the data already collected and if not, what steps are necessary to collect the data?
  - Are additional sensors required?
  - Are the data available in a database? Do the employees need special access for data mining to work with these data?

– What is the quality of the data? Are they complete or do you have to expect missing or incorrect data?

- **Phase 3. Data preprocessing.** Once you understand the data, you can get an idea of how reliable the data are and whether cleaning or selection procedures are necessary or not.
- **Phase 4. Modeling.** This is the practical application of a data mining model, which could mean, for example, applying one of the methods presented in the later chapters on supervised or unsupervised machine learning.
- **Phase 5. Evaluation.** This phase largely corresponds to its abstract analogue in Sect. 1.3.2. Here, the results of the data analysis are critically questioned. An additional aspect in CRISP-DM is the consideration of economic constraints: How well does the data analysis process meet the objectives of a company or production strategic view? The KPIs determined in Phase 1 are used to capture the success and quality of data mining.
- **Phase 6. Deployment.** In this phase, the learned data relationships are practically applied in the company processes. In the process industry, this is the online use during production. This phase brings new difficulties, because now the models gained through data mining must be used to achieve a certain benefit. Either to identify faulty products or process states, or to control the production machines to achieve better quality in advance.

  For deployment, it is important to know how long a model is run and how accurate and precise the model is. Especially when automation components depend on the model output, deployment includes all those tasks to make their operation stable and enable an online application in a closed loop.
- **Phase 7. Monitoring.** A critical evaluation of the performance of data mining models is provided for in the CRISP-DM approach as a separate phase. It is a feedback phase for modeling and allows the process to be corrected and improved iteratively over time.

  However, not only the model quality is monitored. The performance in relation to the KPIs selected in Phase 1 is also taken into account in order to actually quantify the impact of the data mining application (Fig. 1.7).

The CRISP-DM process models the rather abstract data mining tasks described in a model template, with phases 1 and 2 responsible for the essential steps of data mining preparation, while phase 3 includes activities such as data cleaning and transformation. Each phase contains a series of tasks that must be completed, and the results of each phase serve as input for the next phase. The goal of CRISP-DM is to ensure that all necessary steps are taken to create a high-quality data mining model and that the model is used appropriately.

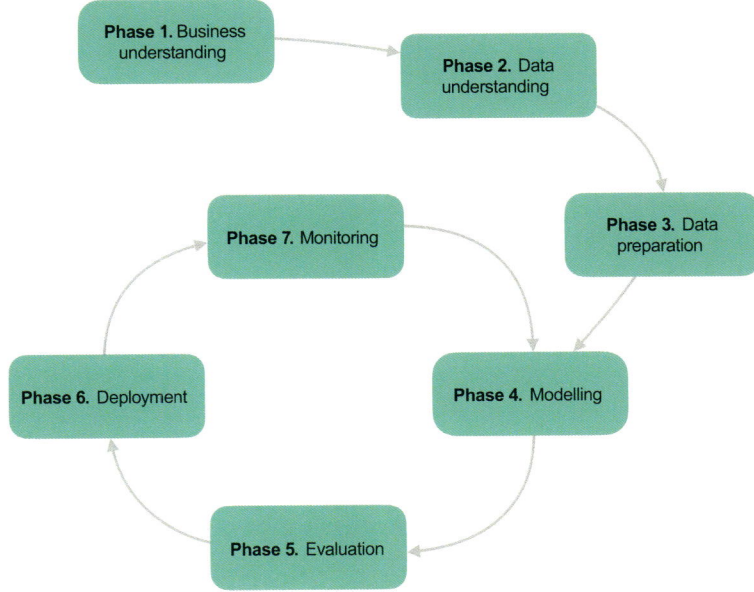

Fig. 1.7   CRISP-DM and its various phases

1.4   Practical Handling of Data in Python

1.4.1   Programs in this Book

Jupyter Notebooks
We will be working with Python throughout this book. All programs will be provided to you as so-called Jupyter Notebooks. In Appendix A, we suggest a suitable Python distribution and the associated programming environment. Appendix A also contains a compilation of basic knowledge in Python. It can help you get to grips with the language and offers an introduction for those who are completely unfamiliar with Python. However, we do assume some basic knowledge of programming languages and computer science for this book.

Why Python?
There are many programming languages that can be used to implement machine learning. Lua, C#, C++, Java, Matlab, and Julia are just a few examples. For applications in the technical field, it is particularly important that a language allows non-computer scientists to get started. Python has many advantages here:

For one thing, Python is easy to learn. Supported by a large community on the net, it is considered one of the best languages for programming beginners. It also has a wide syntactic range. Code can be written functionally or object-oriented, depending on the requirements of a project.

Python is an interpreted language. This means we can execute it directly without performing a compilation. With the help of Jupyter notebooks, we can directly check changes to the code for their effect, which promotes testing and simply playing with the code.

If you want to use your own learning methods in industry, it is easy to integrate solutions developed in Python into existing infrastructures. Almost every data management system of larger companies, manufacturing companies or even authorities offers interfaces to this programming language. It runs on both large computer centers and microcomputers like the Raspberry Pi.

1.4.2   Libraries for Python

If you implement algorithms purely in Python, they are often very slow, but can be formulated in a way that is easy to understand. Where speed matters, Python uses special libraries written in C++ code.

Libraries are therefore the component that makes Python powerful. There is a wealth of scientific tools that are tested, proven, and ready to use. These libraries are free and can be used by anyone. The following list is not complete, but reflects the tools we will use more frequently in this book:

- **Numpy.** Numpy [6] is a framework that provides us with various mathematical operations such as vectors, matrices, the Fourier transformation, and a number of tools for data transformation.
- **Matplotlib.** Matplotlib [7] is the gold standard for creating academic diagrams and visualizations in Python. We use this library for plotting.
- **Pandas.** Pandas [11] is a library dedicated to handling datasets, manipulating datasets, and easy saving and loading of data. It offers us an easy way to load, save, and work with data. With Pandas, extremely fast vector and matrix operations are also possible [8]. However, we limit our use of Pandas somewhat and try to get by with the structures of Numpy where possible.
- **Scikit-Learn.** The Scikit-Learn library [9] is a diverse, easy-to-use toolkit for applying almost all methods we introduce in this book.
- **Tensorflow.** Tensorflow [1] is a library from Google for computing with tensors, as needed for building neural networks, for example.
- **Keras.** Keras [3] provides a simple interface for building and working with neural networks that directly builds on Tensorflow. It simplifies the creation of networks and allows you to quickly create solutions.

1.4.3   Explainable Data in Python

The concept of explainability is at the forefront of many works that use algorithms in practice. With this term, we capture the ability to understand a model or a relationship. How did an algorithm arrive at its result? Why did a computer program make a certain decision? If we can explain the decision-making process in an understandable way, trust in our model increases. This consideration will become important later, especially in the context of learning methods. Are we even able to trust such a method? We will therefore get to know approaches that allow us to look deeper into learning methods and better understand how they work.

Explainability includes a basic understanding of the input variables as well as their significant properties and boundary conditions. In many real database systems, this level of variable description is missing. Sometimes the true meaning of the captured folders is not digitally recorded, at worst only in the head of an employee. This carries enormous risks and greatly complicates practical work.

Units of measurement

A first clear step towards explainability is the consistent consideration of units of measurement. They define the scales on which we measure. Measured variables are a product of their value $\{x\}$ and their unit $[x]$,

$$x = \{x\}[x]. \tag{1.2}$$

Without knowledge of the unit of measurement, the data is practically worthless. When dealing with technical data and process variables, it is therefore crucial to keep a close eye on the units. In some situations, the units of variables can also change over time. Just think of a scale that initially measures in grams and then automatically switches to kilograms for its measuring range when something heavier than 1 kg is placed on it. Consequently, we must always include the units in our data handling. One way to do this is shown in Listing 1.1. Here you see two arrays for t and x as well as a dictionary.

Listing 1.1 Treatment of units

```
1  t = [0,1,2,3] # array for indices
2  x = [1.0,2.0,3.0,4.0] # array for measurement
3  myDictionary = {'Time':t, 'Position':x, '
       Position_Unit':'m', 'Time_Unit':'s'}
```

Here we capture additional information alongside the actual data in the arrays, by including them in the data structure myDictionary and storing them as additional fields. Often it is enough to integrate a description in the form of a string into the dictionary.

Units are a basic prerequisite for the **explainability** of a variable.

Unfortunately, units are completely missing or are only visible through secondary tables in surprisingly many real process databases. An automatic check of the units is then difficult.

During data collection, changes in units should definitely be avoided. They increase the complexity of preprocessing without generating added value. Why are unit jumps observed in datasets at all? The main reason lies in the display of numbers. Digital measuring devices always try to display their values on an optimal scale. However, the data indication optimized for a display is not always optimal for data storage. If you agree on consistency in the unit, you save time and effort in your later evaluation. You also eliminate a known source of error.

Taking uncertainties into account

In the case of measured quantities, the inclusion of uncertainties helps us to further increase our understanding of a variable. It provides us with a quantitative way to describe our confidence in the respective quantity. The absolute uncertainty u_x for any quantity x, would describe the measured value x_m in the following way,

$$x_m = x \pm u_x. \tag{1.3}$$

The absolute uncertainty is related to a relative uncertainty by

$$\varepsilon_x = \frac{u_x}{x_m}. \tag{1.4}$$

These two uncertainties can be used in our dictionary to describe the data, as shown in the following extension of our code example:

Listing 1.2 Example for adding stochastic variables

```
t = [0,1,2,3] # array for indices
x = [1,2,3,4] # array for measurement
myDictionary = {'Time':t, 'Position':x, '
    Position_Unit':'m', 'Time_Unit':'s', '
    Position_Absolute_Uncertainty':0.05, '
    Time_Absolute_Uncertainty':0.01}
```

Knowing uncertainties, communicating them to the user, and ultimately always considering them in the larger context, is a first step in building trust in an algorithm.

> The informed indication of uncertainties increases the **trust** in data. Agree on a form to indicate these uncertainties, absolute or relative, and consistently maintain this form for your data collection.

Summary

In this first chapter, we have dealt with data. Our goal is to create models from data using machine learning. With these models, we can analyze and improve technical processes. To do this, we first had to show ways in which we can describe, characterize, and structure data. We have seen that in addition to fully structured and unstructured data, there is also a semi-structured type.

Furthermore, first, simple steps in Python were introduced. Here we have seen that the systematic consideration of units and the treatment of uncertainties play a major role in data evaluations. Without knowledge of their unit, a physical quantity is incomprehensible. Without knowledge of the uncertainty, we do not know how far we can trust the data value.

Tasks

1.1 Characterize the following data collections!

- Phone book
- Calendar entries
- Address book for contacts
- Electricity consumption

1.2 Why is focusing important in data mining? Can you provide reasons why not all possible data is always considered?

1.3 What type of scale is present in the following data sets?

- Scale for measuring length
- Rowing motion in an airplane
- Pressure measurement in hPa
- Collection of fish names
- Wood colors
- Camera images

1.4 Let's assume you have measured rolling forces in an aluminum plant. What additional information do you need in order to draw conclusions about the quality from the data?

1.5 You work in a sawmill and your saws wear out over time. a) What variables would you record to trace the cause? b) How would you measure the variables? c) What types of data and scales are involved with your data?

References

1. M. Abadi, A. Agarwal, P. Barham, E. Brevdo, Z. Chen, C. Citro, G. S. Corrado, A. Davis, J. Dean, M. Devin, S. Ghemawat, I. Goodfellow, A. Harp, G. Irving, M. Isard, Y. Jia, R. Jozefowicz, L. Kaiser, M. Kudlur, J. Levenberg, D. Mané, R. Monga, S. Moore, D. Murray, C. Olah, M. Schuster, J. Shlens, B. Steiner, I. Sutskever, K. Talwar, P. Tucker, V. Vanhoucke, V. Vasudevan, F. Viégas, O. Vinyals, P. Warden, M. Wattenberg, M. Wicke, Y. Yu, and X. Zheng. TensorFlow: Large-scale machine learning on heterogeneous systems, 2015. https://www.tensorflow.org/. Software available from tensorflow.org.
2. S. Aggarwal and N. Manuel. Big data analytics should be driven by business needs, not technology. *McKinsey & Co.*, June 2016.

3. F. Chollet et al. Keras, 2015. https://github.com/fchollet/keras.
4. L. Fahrmeir, R. Künstler, I. Pigeot, and G. Tutz. *Statistik*. Springer, 1997.
5. J. Hamilton. *Time Series Analysis*. 1992.
6. C. R. Harris, K. J. Millman, S. J. van der Walt, R. Gommers, P. Virtanen, D. Cournapeau, E. Wieser, J. Taylor, S. Berg, N. J. Smith, R. Kern, M. Picus, S. Hoyer, M. H. van Kerkwijk, M. Brett, A. Haldane, J. F. del Río, M. Wiebe, P. Peterson, P. Gérard-Marchant, K. Sheppard, T. Reddy, W. Weckesser, H. Abbasi, C. Gohlke, and T. E. Oliphant. Array programming with NumPy. *Nature*, 585(7825): 357–362, Sept. 2020. https://doi.org/10.1038/s41586-020-2649-2.
7. J. D. Hunter. Matplotlib: A 2d graphics environment. *Computing in Science & Engineering*, 9(3): 90–95, 2007. https://doi.org/10.1109/MCSE.2007.55.
8. T. pandas development team. pandas-dev/pandas: Pandas, Feb. 2020. https://doi.org/10.5281/zenodo.3509134.
9. F. Pedregosa, G. Varoquaux, A. Gramfort, V. Michel, B. Thirion, O. Grisel, M. Blondel, P. Prettenhofer, R. Weiss, V. Dubourg, J. Vanderplas, A. Passos, D. Cournapeau, M. Brucher, M. Perrot, and E. Duchesnay. Scikit-learn: Machine learning in Python. *Journal of Machine Learning Research*, 12: 2825–2830, 2011.
10. C. Shearer. The crisp-dm model: The new blueprint for data mining. *Journal of Data Warehousing*, 5(4), 2000.
11. Wes McKinney. Data Structures for Statistical Computing in Python. In Stéfan van der Walt and Jarrod Millman, editors, *Proceedings of the 9th Python in Science Conference*, pages 56–61, 2010. https://doi.org/10.25080/Majora-92bf1922-00a.

Chapter 2
Mathematical Description of Data

Keywords Stochastics · Statistics · Distribution functions · Bayesian statistics · Modeling of uncertainty

Understanding many machine learning methods requires knowledge of statistics. Since the treatment of uncertainties in data is a central element of the book, an introduction to stochastics is provided first. The reader is guided through a review of set theory to the definition of probability and conditional probability. Relevant tools from statistics, such as the calculation of expected value, variance, covariance, correlation, and distributions, are explained.

The mathematical discipline of stochastics deals with the concept of probability. It is capable of modeling processes that are completely or partially influenced by chance. Throughout this chapter, we will review important aspects of stochastics and use it to view data as stochastic processes. After that, we will use statistical tools to extract information from the data.

Why does this matter for machine learning? We have listed an overview in Fig. 2.1 that shows which concepts the understanding of this chapter is relevant for. Primarily, our goal is to build robust, explainable algorithms. These algorithms must take into account the influence of chance in data and be able to model it as well. Machine learning and statistical analysis have a thematic overlap. They merge seamlessly into each other. For approaches of physics-informed learning, statistical properties of data series, analytical model equations, and semantic relationships are used to provide additional information to learning processes. Knowledge about uncertainties and their statistical distribution can be such additional information. Statistical concentration measures and entropy are used by learning methods to distinguish data influences. They determine how important the information in a variable is and how much influence it can have on a result.

M. J. Neuer, *Machine Learning for Engineers*,
https://doi.org/10.1007/978-3-662-69995-9_2

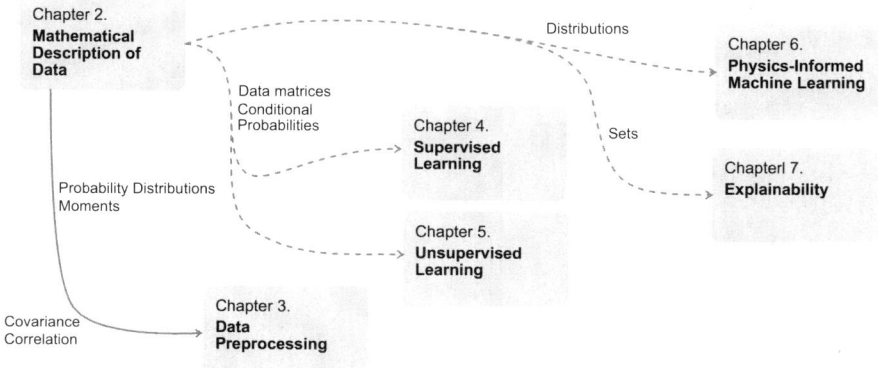

Fig. 2.1 Overview of how this chapter influences the subsequent chapters

After the basics of stochastics, we turn to statistical tools. However, we will not discuss these in full detail. Instead, we restrict ourselves to concepts that we often need for machine learning, and to those mathematical relationships, which are absolutely necessary for the learning processes.

In the area of mathematical description, there are more complete, closed considerations. Here we recommend the book by Van Kampen [17] for a comprehensive presentation of stochastic processes, and the books by Fahrmeir et al. [2] and Toutenburg et al. [15] for the detailed treatment of statistical tools. If it is didactically sensible, we deviate slightly, but purposefully, from the usual nomenclature in the statistical literature.

2.1 Basics of Stochastics

2.1.1 Probability

Probability is the ratio of the number of favorable events to the total number of all events. This important relationship is the probability definition of Laplace. It is a concept that is intuitively connected with our daily life, and often finds application there. Many methods, such as Bayesian networks and decision trees, are based on this stochastic perspective.

> **Example: Parking Search**
> The ratio of free parking spaces to the total number of all parking spaces tells us how likely we are to find a free space. Even if we do not consciously perform this calculation, it takes place unconsciously and immediately whenever we search for parking spaces. ◀

2.1.2 Sets

The set theory is a fundamental area of mathematics. It is treated in many books, for a more detailed discussion, the book by H.-D. Ebbinghaus [1] is recommended. For us, it represents a starting point. Understanding sets helps us in formalizing relationships such as "is part of" or "belongs to the group" and therefore offers a very natural way of semantic modeling.

> **Example: Football Team**
> Imagine you are part of a football team T and are identified by a name x, then we formalize this mathematically as $x \in T$ and say, x is an **element** of the **set** T. Let's further assume that you enjoy listening to a famous band and therefore belong to their fan club Q, then also $x \in Q$ is true. ◀

Since you belong to both groups, there is a relationship between these two sets. They have at least one common element—you—and we can say that T and Q have an overlap, which we write as follows,

$$C = T \cap Q. \tag{2.1}$$

It is referred to as the **intersection** or **conjunction** of T and Q. There may be further $x \in (T \cap Q)$ and all these elements form a new set, for whose elements

$$x \in (T \cap Q) \Leftrightarrow x \in T \wedge x \in Q \tag{2.2}$$

must apply. Please keep in mind that not every $x \in T$ is automatically also element in Q and vice versa. Rather, the condition is that every element of $x \in (T \cap Q)$ must necessarily be in both—represented by the mathematical "and" symbol ($\wedge$) in (2.2).

In contrast, we can form the **common set** or **union set** by merging all elements of both sets T and Q,

$$S = T \cup Q. \tag{2.3}$$

Then $x \in T \cup Q$ means that x can be a member of T or a member of Q or both—which is what the mathematical "or" $\vee$ states,

$$x \in (T \cup Q) \Leftrightarrow x \in T \vee x \in Q. \tag{2.4}$$

We also say that this common set S is a **superset** of T and Q, while T and Q are **subsets** of S. Of course, a precise understanding of a set and its elements is essential for any subsequent analytical treatment.

A final relevant term from set theory describes sets that share no elements at all. We then say the sets are **disjoint** and it applies

$$x \in T \Leftrightarrow x \notin Q \text{ und } y \in Q \Leftrightarrow y \notin T \tag{2.5}$$

as well as

$$T \cap Q = \emptyset. \tag{2.6}$$

The empty set $\emptyset$ is, by the way, disjoint with all sets.

Often we need to know how many elements a set has. The **cardinality** M of a set Q indicates this number of its elements and is abbreviated with $M = |Q|$.

2.1.3 Probability Definition According to Laplace

Let us begin this section with an example that is synonymous with the concept of probability like no other: the dice.

Example: Dice

Imagine a regular six-sided dice. Intuitively, you assume that the probability of rolling a 4 is 1/6, as you have six possible outcomes and only one case, namely 4, is in your favor. The probability of getting only odd numbers, here 1, 3, 5, is similarly given by $3/6 = 1/2$. In set notation, all possible sides would be captured in a set $\Omega = \{1, 2, 3, 4, 5, 6\}$, with the cardinality $|\Omega| = 6$. The set $A = \{1, 3, 5\}$ has the cardinality $|A| = 3$. ◀

This form of calculating probability was developed by Laplace and is one of the foundations of stochastics. The general definition of probability is:

Probability According to Laplace Let A be a set of events and Ω a set that contains A, $\Omega \supset A$. The size $P(A)$, given as

$$P(A) = \frac{|A|}{|\Omega|} = \frac{\text{all favourable events}}{\text{all possible events}}, \tag{2.7}$$

then indicates the probability of the occurrence of A.

This statement assumes that we understand what an event is in the first place. How can we count in sets that contain events? How can we formalize the establishment of a probability measure given by (2.7)?

2.1.4 Event, Outcome, and Probability Spaces

In the definition (2.7) of $P(A)$ we need two special sets. They help us describe events and outcomes. The outcome set Ω contains all possible (allowed) outcomes. In the case of the dice in example 3 this is

$$\Omega = \{1, 2, 3, 4, 5, 6\} \tag{2.8}$$

An event A is a subset of $\Omega \supset A$, e.g. the odd numbers of the dice

$$A = \{1, 3, 5\}, \tag{2.9}$$

which contains three favorable outcomes. In fact, we say an event has occurred if even just one favorable element of the event A was rolled. Please note the difference between the elements of A and A itself. In the present case, A has according to (2.7) the probability $P(A) = 3/6 = 1/2$ to be true.

We summarize all possible events A_i in the event space $\Sigma = \{A_1, A_2, \ldots\}$. The event space Σ is thus the set of all subsets of Ω. Each A_i lies in Σ. The example with the odd numbers also shows that there are events that contain several result elements. If an event A^* contains only a single result, e.g. $A^* = \{3\}$, we call A^* an **elementary event.**

There are two important, special events that we would like to highlight. First, the entire result set Ω is also an event set—it reflects the **certain** event. In the dice example, the result will always be a number between 1 and 6. On the other hand, the (abstract) **impossible** event describes a case that can never occur. Thus, the empty set $A = \emptyset$ is an impossible result.

With the result space Ω and the event space Σ we set up another space, the so-called probability space Π:

> **Probability Space** We call $\Pi = (\Omega, \Sigma, P)$ a probability space with the outcome space Ω and the event space Σ, if there is a mapping rule that assigns a measure $P(A)$ to each event element $A \in \Sigma$ that quantifies the probability of A occurring.

In mathematics, this process is not trivial. After all, they must algebraically ensure that the assignment from the two sets Ω and Σ to such a measure $P(A)$ is possible at all.

2.1.5 Axioms of Kolmogorov

To be able to deal with probabilities, one needs a set of rules. These rules are given by the axioms of Kolmogorov [9]. They help us to determine $P(A)$ and to calculate with it.

First Axiom of Kolmogorov The probability $P(A)$ is a number between 0 and 1,

$$0 \leq P(A) \leq 1. \tag{2.10}$$

This axiom limits the values of $P(A)$. According to this convention, a probability can never be greater than 1 or less than 0. Of course, you are free to come up with your own scale for $P(A)$, a scale that sets other limits. But you would always be forced to agree on the two boundary cases of the certain event and the impossible event. If we look again at these special events from Sect. 2.1.4, the first axiom with its choice of limiting $P(A)$ directly leads to being able to assign concrete numerical values to these events:

Second Axiom of Kolmogorov The certain event has a probability of 1, $P(\Omega) = 1$ and the impossible event has a probability of 0, $P(\emptyset) = 0$.

The definition in the first axiom and the choice of the limits 0 and 1 is therefore sensible. Ultimately, we need a calculation rule for probabilities and this is provided by the third axiom:

Third Axiom of Kolmogorov If two event sets are disjoint, $A \cap B = 0$, then applies

$$P(A \cup B) = P(A) + P(B). \tag{2.11}$$

From the axioms, we can derive some conclusions that greatly simplify the handling of probabilities. First of all, elementary events represent the ideal case of disjoint events. They are therefore always covered by the third axiom. The inverse event $\neg A$ refers to all events that are not A, $\neg A = \Omega \setminus A$. Because of the second axiom, the following applies:

$$P(A \cup \neg A) = 1 \quad \Leftrightarrow \quad P(\neg A) = 1 - P(A), \tag{2.12}$$

This statement is a consequence of the axioms, which we often use for the calculation of probabilities.

Example: Rolling the inverse set
The probability of rolling a 3 is $P(3) = 1/6$. Because of (2.12) is $P(\neg 3) = 5/6$. ◀

Fig. 2.2 Illustration of overlapping event sets

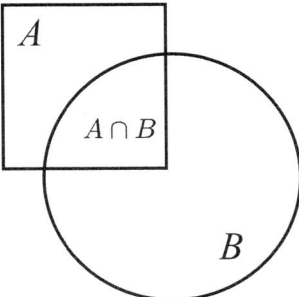

The third axiom in (2.11) can be extended to any event sets. For this, we consider Fig. 2.2. The probabilities cannot be added arbitrarily here, since A and B have a common intersection. This intersection would simply be counted twice when determining the probability. The events of $A \cap B$ are taken into account in $P(A)$ and in $P(B)$. If you are aware of this, then you can also directly subtract this partial probability. In this way, we reach the addition theorem for probabilities:

Addition theorem for probabilities For any two event sets, the following applies:

$$P(A \cup B) = P(A) + P(B) - P(A \cap B). \tag{2.13}$$

While the last equation helps us to add probabilities, probabilities can also be multiplied. If event A is expected first, it has the probability of occurrence $P(A)$. If event B is to occur afterwards and this event is in no case influenced by A, i.e., statistically independent of A, then the probabilities of A and B can be multiplied to calculate the joint probability for the occurrence of both events:

Multiplication of Probabilities For statistically independent event sets A and B, the following applies

$$P(A \cap B) = P(A)P(B). \tag{2.14}$$

2.1.6 Conditional Probability and Bayes' Theorem

After we have a sufficiently good understanding of the concept of probability, it is interesting to delve a little deeper into the dependencies of probabilities. So we ask the question, how likely is the occurrence of event A if event B has already occurred.

Conditional Probability If A and B are two event sets, then $P(A|B)$ describes the probability of the event of A if the event B has already occurred. $P(A|B)$ is called a conditional probability and it applies,

$$P(A|B) = \frac{P(A \cap B)}{P(B)} \qquad (2.15)$$

In Fig. 2.3, an outcome space is outlined that contains several event sets A, B, C_1 and C_2. Using this drawing, we can more easily consider the abstract definition of conditional probability. It is immediately apparent that C_1 is a subset of A. Here, $P(A|C_1) = 1$ must necessarily apply, as with every element of C_1 that occurs, the entire event set A has also occurred. Remember that the concept of an event set occurring means that only a single element actually becomes true. Also, $P(A|C_2)$ is easy to determine. No element of C_2 overlaps with A. Consequently, A cannot occur under the condition that C_2 has already occurred: $P(A|C_2) = 0$.

The most interesting case in Fig. 2.3 is, however, B. Here, a part lies in A and another part outside of A. The conditional probabilities for $P(A|B)$ and $P(B|A)$, depend on each other in the sense of Definition 2.1.6

$$P(A|B)\,P(B) = P(A \cap B) = P(B|A)\,P(A), \qquad (2.16)$$

which forms the basis for Bayes' theorem:

Bayes' Theorem For the conditional probabilities $P(A|B)$ and $P(B|A)$ applies

$$P(A|B) = \frac{P(B|A)P(A)}{P(B)}. \qquad (2.17)$$

The significance of conditional probability can be illustrated by the following examples:

Fig. 2.3 Several event sets with their partial probabilities (green)

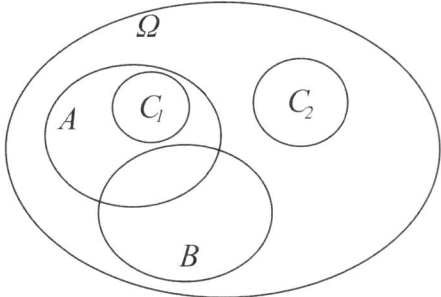

Example: News channel survey as conditional probability
A survey is being conducted on a news channel. Viewers are asked to vote
on their opinion on a particular topic. They can choose between answer A
and answer B. As a result, the viewer receives distribution values, which proportion voted for A and which proportion voted for B. However, what you
see here is only the part of viewers who i) could watch this channel at the
time of the surveys and ii) also take the initiative to call in on the topic. ◄

Example: Neighborhood of two points
The neighborhood of two data points can be modeled using conditional
probabilities. If two points are close to each other, this probability should be
high. At larger distances, it should be low. A simple model for this is

$$p(i|j) \propto \exp\left(\frac{-(x_i - x_j)^2}{2\sigma^2}\right). \tag{2.18}$$

◄

Example: Conditional Events
Two   discrete   vectors $a = [0,0,1,1,0,1,0,1,1,0]$ and
$b = [0,0,0,1,0,1,0,0,1,0]$ each represent 10 simultaneous events. The
first entry in a corresponds to the first entry in b and so on. The conditional
probability for $b = 1$ given that $a = 1$ has already occurred is given by
$p(b = 1|a = 1) = 3/5$, because in 5 possible cases $a = 1$ occurs, and in 3
favorable cases $b = 1$ occurs simultaneously. ◄

2.1.7 Stochastic Process

With the previous concepts of event and outcome set as well as the concept of probability P, we will next describe data variables that can be influenced by chance as
so-called stochastic processes (or random variables). In his book on stochastic processes [17], Van Kampen extensively shows several examples from physics and
chemistry as well as the mathematical foundation to deal with chance in nature. We
start here with a discrete random variable as given, for example, in dice rolling.

A variable x is called a stochastic process if its values depend on a random pro-
cess in the event space Σ that is defined by $P(A)$ with $A \in \Sigma$. A realization $x_r \in \mathbb{R}$
occurs when the random variable has the specific value $X = x$, the occurrence of
which is captured by the probability $P(x = x_r)$. Let's illustrate discrete random
variables with some examples:

Example: Coin Toss
The variable x is the result of a coin toss. We define that $x = 0$ when heads is tossed and $X = 1$ when tails is up. The probability for $x = 0$, thus $P(x = 0) = 0.5$, also describes the behavior of the variable. Here, $x_{r,1} = 0$ and $x_{r,2} = 1$ are the realizations of x, as described in the definition. ◀

When we assign values $P(A)$ to events A_i in the probability space, the question arises as to how probabilities can be arranged over a continuous event space. With this knowledge, we would see at a glance whether there are particularly likely events and where they are located. While the assignment of a probability in the discrete case can be done very directly, we have to consider an interval for continuous functions. The probability now runs not in steps, but as a continuous function:

Probability Density The probability $P(\alpha \leq x \leq \beta)$ of a continuous stochastic process is given by

$$P(\alpha \leq x \leq \beta) = \int_{\alpha}^{\beta} p(x)dx \qquad (2.19)$$

and $p(x)$ is called the probability density of x.

Understanding stochastic processes is very important for the concept of probability density. As a function, it captures the nature of probability and gives us a tool to generate any random processes ourselves.

Since we demand in the first axiom of Kolmogorov that the probability should lie between 0 and 1, the probability density $p(x)$ must be normalized accordingly. Its entire area must reflect the certain event,

$$P(\Omega) = \int_{-\infty}^{\infty} p(x)dx \overset{!}{=} 1. \qquad (2.20)$$

This does not mean that $p(x)$ is less than 1 at every point. $p(x)$ can take any value as long as the normalization strictly applies. For this reason, you should not confuse probability densities with probabilities.

If a variable is composed of a deterministic and a stochastic component, we write

$$x = x_D + u_x \qquad (2.21)$$

and describe with x_D the deterministic part and with u_x the stochastic process.

Example: Temperature fluctuation by 5 °C
The variable T is a stochastic process. Its values fluctuate $\pm 5°C$ around a value of 20 °C. If you write $T = \tilde{T} + u_T$, you separate the deterministic part from the stochastic process. In this case, $\tilde{T} = 20\,°C$. ◀

We will also call the probability density $p(x)$ probability distribution or simply distribution. On the other hand, we call a function F, for which

$$F(x') = \int_0^{x'} p(x)dx, \tag{2.22}$$

is called **cumulative distribution function** of the probability distribution $p(x)$.

2.1.8 Considering the Probability Density in the Data

We have already seen in Chap. 1 how we can incorporate units and uncertainties into our work with Python. But not only the uncertainty plays a role in the description of a data series. The probability distribution from which the data series originates also contains valuable information.

Listing 2.1 Continuation of Listing 1.2, adding semantic descriptions for the probability density

```
myDictionary = {'Time':t, 'Position':x, 'Position_Unit':'m', '
    Time_Unit':'s', 'Position_Absolute_Uncertainty':0.05, '
    Time_Absolute_Uncertainty':0.01,                           '
    Position_Probability_Density':'Pearson_IV'}
```

Listing 2.1 adds this based on a `string`, specifying a Pearson-IV function as the probability distribution. Of course, it is also possible to store a Python function here that you have defined accordingly. Then, for example, a piece of code that is embedded in the string would result for the entry in the dictionary:

Listing 2.2 Analytical probability density as text in a data object

```
# [... siehe oben ...]
'Position_Probability_Density':'A*exp(-(x-mu)**2/2/Sigma**2',
    ...
```

This form of data storage represents a Python-based way to meet the requirements of the digital system of units (D-SI) as specified in [7]. Current work, such as that by B. D. Hall and M. Kuster [3], suggests that the consideration of the scale is also important, which we will briefly implement in the next listing.

Listing 2.3 Integration of a specific type of scale

```
# [... siehe oben ...]
'Scale':'Cardinal',
'RangeMax':0.4, 'RangeMin':0.1,...
```

2.2   Stochastics of Data

How do chance, uncertainty, and probability relate to our data? In the previous chapter, we briefly explained how measurement uncertainties are formally integrated into a data approach. With the concepts of stochastics, we can formally describe the actual process of data collection. We want to go into more detail on this in the following.

2.2.1   Stochastics of the Sampling Process

Any abstract continuous data series $\xi(t)$ can be digitally sampled by performing a measurement at certain points in time τ_i. i numbers these measurements and is represented by a natural number $i \in \mathbb{N}_0$. In this book, we always use the natural numbers with zero, $\mathbb{N}_0 = (0, 1, 2, 3, 4, \ldots)$. Zero-based indexing is common in programming and aids in the understanding of algorithms.

> **Sampling**   The data points x_i of a data series are determined by measurements
>
> $$x_i = \xi[\tau_i + \delta t)] + u_x \tag{2.23}$$

at defined points in time τ_i, where the $\tau_i \leq \tau_{i+1}$ are referred to as **sampling points**. When all intervals τ_i have the same distance Δt, we say the data points are **equidistant** in time with $\tau_i = i\Delta t$. The sampled signal x_i is then determined by

$$x_i = \xi[i\Delta t + \delta t] + u_x, \tag{2.24}$$

where the time Δt is referred to as the **sample time** and $\nu = \Delta t^{-1}$ as the **sampling frequency**.

In this definition, i indicates the index of a data point and u_x is the uncertainty of the measurement x_i of $\xi(i\Delta t)$. We assume that ξ is a kind of basic truth or in other words: ξ is the process itself. We will never know ξ exactly, because our measurement method is always associated with the **absolute uncertainty** u_x. This uncertainty is a random number that affects every measurement, as sensors and measurements only have limited accuracy. The absolute uncertainty u_x can also be stated as **relative uncertainty** $\varepsilon_x = u_x/x$.

Another stochastic influence, which is also explicitly represented in (2.24), is the uncertainty of our clock. It is described by a jitter δt on $\Delta t = \Delta t_0 + \delta t$. This is only mentioned here for the sake of completeness, as it will not play a role in our later considerations. In Fig. 2.4 we illustrate the difference between the real process $\xi(t)$ (green), the measurement points x_i (black) and the measurement uncertainty ε (red). Because u_x and δt are stochastic processes, x_i is a stochastic process.

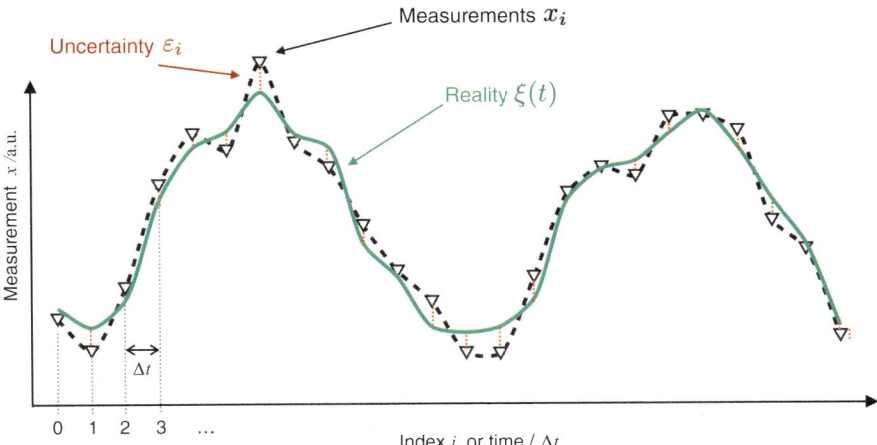

Fig. 2.4 Real values of a process variable (green), measured data points (black triangles) and an example of measurement uncertainty (red)

2.2.2 The Nyquist Theorem

A key quality feature of our sampling is the frequency with which we observe and measure the process, essentially determined by Δt, which results in the **sampling frequency** or **sampling rate**.

$$\nu_S = \frac{1}{\Delta t} \tag{2.25}$$

of the data points x_i. It is obvious that the number of data points used for sampling the behavior of a variable is important to capture the entire amount of information. This becomes clear when considering a concise example, such as the movement of a pendulum: If you only measure whenever the pendulum crosses its zero point, you will find that according to your data there is no movement at all. In this sense, the dynamics of the system were not captured by the measurement, simply because the sampling frequency was poorly set.

In signal theory, the Nyquist-Shannon theorem is well-known [14]. If you want to fully capture the dynamics of a system, you must sample at least twice as fast as the highest frequency you observe in the system,

$$\nu_S \geq 2\nu_{max}. \tag{2.26}$$

Conversely, the highest resolvable frequency is obtained by the sampling frequency. This highest frequency is defined as the **Nyquist frequency**.

A digital signal, which is sampled with ν_S, can resolve dynamic components up to a maximum of

$$\nu_{\text{Nyquist}} = \frac{1}{2}\nu_S, \tag{2.27}$$

. This frequency is called the **Nyquist frequency**.

In practice, however, it is better to know the dynamics of the measured system in detail and make an appropriate estimate for ν_S. For example, consider a heating process of a piece of steel in a furnace that lasts several hours. In this case, the temperature change is only interesting on a scale of minutes—not seconds. In a fast process, such as the rolling of flat steel, sampling times in the sub-second range are common.

Recent work on information theory, such as the work by Vetterli et al. [18], explains the Nyquist theorem in connection with the so-called **rate of innovation** (engl. *finite rate of innovation*). They found out that the degree of uncertainty of a signal plays the decisive role and not necessarily just the frequency. This can be easily illustrated with the following example: If you sample a sinus and have received information about its approximate frequency in advance, only a few sample points are sufficient for sampling. So if certain information about the signal of the measured variable is known in advance (engl. *prior knowledge*), lower sampling frequencies are sufficient to fully capture the information.

2.2.3 Distinguishing Uncertainties

Aleatory Uncertainty

Example: Lottery as Aleatory Uncertainty
Assume you play the lottery regularly. If you are asked whether you will win in this lottery the next day, you cannot make a reliable statement. You do not know because it purely depends on chance. At most, you know the probability of winning. ◄

You cannot predict the occurrence of this event. This form of inherent randomness is called **aleatory uncertainty** (lat. *alea,* the dice). No matter how often you play, it will never disappear. Examples of this are:

- Weather phenomena such as wind or rain,
- Earthquakes and volcanic eruptions,
- Movement of Saturn's moon Hyperion (which is actually not deterministic),
- Coin toss,
- Radioactive decay.

For many of these processes, models exist that contain the stochastic components. The random component does not always dominate. In fact, there are many situations where we can describe the randomness of the process as pure disturbance. The model then contains a deterministic part and a stochastic part. In Monte Carlo simulations, such models can finally be simulated and evaluated using statistical tools.

This term is also known in other areas. In music, for example, the use of surprising motifs and elements is referred to as aleatory composition.

Epistemic Uncertainty

Uncertainty also arises from too few data. Averages and variances have little significance when calculated on very small data series. The computational accuracy of a processor and the mathematical depth of the model are such uncertainties. In theory, however, all these factors can be avoided, either by more computing power, more data points, better modeling, or exclusion of further systematic deviations. Such uncertainty is called epistemic. It is opposed to aleatory uncertainty.

Relevance of aleatory and epistemic uncertainty in technical applications

S. Maskell shows in [10] how important the consideration of different uncertainties is for information evaluation. Especially when you want to decide in conflict cases between several data sources, you need to know the reliability of the sources. Such considerations also underlie data and sensor fusion approaches, see S. J. Julier et al. [8] and C. Henke et al. [4].

In a paper on mobile sensor technology, I. Yadav and H. G. Tanner present a control concept with stochastic differential equations. For fusion plasmas, stochastic concepts enable the prediction of particle transport, as shown by E. Vanden-Eijnden in [16].

The stochastic perspective also helps in dealing with hazardous substances. S. Hora considers in [5] probability statements about hazardous goods with regard to the question of what influence aleatory and epistemic uncertainties have in this area. Hüllermeier and Waegeman go more fundamentally into the consideration of uncertainty in machine learning methods in [6].

2.3   Statistical Tools for Handling Data

2.3.1   Vector Representation of a Single Data Series

Often it is helpful to represent data as a vector. For simplicity's sake, we assume that we are considering a time series of the variable x. It consists of M measurements of the variable x_i, numbered with the index i from 0 to $M - 1$. The x_i are ultimately combined into the vector x

$$x = (x_0, x_1, x_2, \ldots, x_{M-1}) \tag{2.28}$$

Our notation again uses indices that start at 0, which helps when transferring to a programming language (where 0-based indexing is also common). Even though we are explicitly talking about a time series here, you can represent any type of data in the above vector form. We use bold letters to denote a vector.

Each individual measurement x_i is, as already indicated above, a stochastic process and thus also the vector x.

2.3.2   Expected Value / Expectation Value

A significant part of data evaluation consists of extracting those aspects of the information that are relevant to us. Therefore, statistical key features are of great importance for understanding data series. The most basic property of a group of data is its expected value.

The **expected value or the expectation value** of the data series x with a probability distribution $p(x)$ is defined as

$$\mu = \langle x \rangle_P = E(x) = \sum_{i=0}^{M-1} x_i p(x_i). \tag{2.29}$$

From this general definition of the expected value, one can also easily arrive at the **arithmetic mean**,

$$\mu = \langle x \rangle_{P=1/M} = \frac{1}{M} \sum_{i=0}^{M-1} x_i, \tag{2.30}$$

as soon as one assumes a uniform distribution for p with $p = 1/M$.

Furthermore, we use the notation $\langle . \rangle_P$ for the expected value. In physics, this is often called bra-ket notation. Why is this notation helpful? Please be aware again of the difference between the expected value (2.29) and the arithmetic mean (2.30). In (2.29), an arbitrary probability distribution is assumed. So you are still free to choose this. For different distributions P, we get different expected values. The arithmetic mean (2.30) is a concrete realization of the expected value for the assumption of uniform distribution. We have formally indicated this by a marking $\langle . \rangle_{P=1/M}$ in (2.30). In cases where the reference to the distribution P is already clear, we will omit the designation P at the bracket $\langle . \rangle_P$.

The discrete form of the expected value presented above is applied to digital data. However, at this point, the continuous form of the expected value should also be introduced:

For continuous data, the expected value of x is given by

$$\mu = \langle x \rangle = E(x) = \int_{-\infty}^{\infty} x\, p(x) dx, \qquad (2.31)$$

where $p(x)$ is the probability density of the stochastic process x.

2.3.3   Variance

Let's now turn to the variance. It describes the fluctuation of the data around their average value. Thus, it is a measure to quantify the volatility of a variable x and can be calculated as follows:

The expected value for the square deviation of a measured value x from its expected value is called **variance** and is formally defined as

$$\mathrm{Var}(x) = \sigma^2 = \langle (x - \langle x \rangle)^2 \rangle. \qquad (2.32)$$

So we square the distances between each data point in x and the average value $\langle x \rangle$, then we sum up the squares. In Fig. 2.5 this is illustrated. If the squares are small,

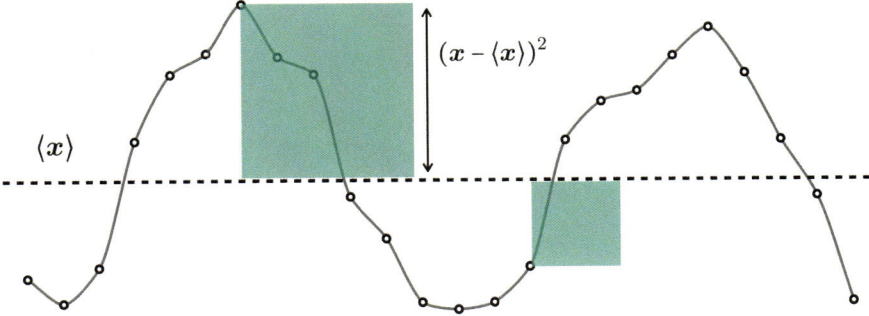

Fig. 2.5 Illustration of variance

there are no large fluctuations in the data and the value remains close to the average value. In contrast, large squares correspond to higher fluctuations and more frequent deviations from the mean.

In the case of variance, the probability p is included twice, as two expected values are also calculated. If you further expand (2.32),

$$\sigma^2 = \langle (x - \langle x \rangle)^2 \rangle = \sum_{i=0}^{M-1} (x_i - \mu)^2 p[(x_i - \mu)^2], \qquad (2.33)$$

you will find, similar to the previous section, the well-known formula for variance with uniformly distributed probabilities $p = 1/M$,

$$\sigma^2 = \frac{1}{M} \sum_{i=0}^{M-1} (x_i - \mu)^2. \qquad (2.34)$$

Please note that the variance is always positive due to the square in (2.34). The only way a negative number could occur here would be through the use of complex numbers to describe data, which, however, is extremely rare. For these cases (e.g., in signal and filter theory, but also in electrodynamics), there are dedicated approaches to determine the variance.

2.3.4 Empirical Variance

The probability distribution of our data is often not known at all. An exact calculation of (2.33) is then strictly speaking not possible. If you repeat a measurement in m processes, there are m different possible realizations for a variable x. The empirical variance is defined in such a situation as

$$\sigma^2_{\text{Empirisch}} = \frac{1}{m-1} \sum_{i=0}^{m-1} (x_i - \mu)^2. \qquad (2.35)$$

In the context of measurement evaluations, studies, and statistical analyses, the empirical variance is often used.

2.3.5 Moments of a Probability Distribution

In the last two sections, we discussed the expected value and the variance. Both quantities are characteristic of the underlying probability distribution of the data. Generally, the so-called i-th moment of a distribution $p(x)$ is defined as

$$m_i = \int_{-\infty}^{\infty} x^i p(x) dx. \qquad (2.36)$$

From the moments of a distribution, we derive the so-called central moments from, by referring them to the expected value over $x \mapsto x - \langle x \rangle$,

$$\widetilde{m}_i = \langle (x - \langle x \rangle)^i \rangle = \int_{-\infty}^{\infty} (x - \langle x \rangle)^i p(x)dx. \tag{2.37}$$

The central moments identify properties of the distribution that are of interest to us. The first central moment is $\widetilde{m}_1 = \langle x - \langle x \rangle \rangle = \langle x \rangle - \langle x \rangle = 0$. By comparing (2.37) for $i = 2$ with (2.32), it can be seen that the second central moment is the variance.

The third moment,

$$v(x) = \langle (x - \langle x \rangle)^3 \rangle, \tag{2.38}$$

is referred to as skewness or skewness (engl. *skewness*). It indicates how strongly the distribution is tilted to one side. The fourth moment is called kurtosis (engl. *kurtosis*)

$$\kappa(x) = \langle (x - \langle x \rangle)^4 \rangle \tag{2.39}$$

and indicates whether the distribution has a flattening at the maximum or tapers sharply. There are distributions whose higher moments help to describe asymmetric behaviors. Around 1900, Karl Pearson extensively dealt with such distributions in [11, 12] and [13]. In the literature, it is common to normalize both the third and the fourth central moment with the variance.

2.3.6 Selected Probability Distributions

Some frequently occurring distributions can be found in the literature. At this point, we will discuss some interesting functions that are important for many technical applications.

If we can determine a probability distribution for data variables, then we can a) use the moments to make statements about the data, b) we can identify individual data points as normal or exotic, and c) we can possibly use the moments to reduce the dimension. To do this, it is necessary to briefly explain some selected distributions.

Gaussian distribution, Normal distribution

The most well-known distribution function is certainly the Gaussian distribution (*gaussian*). This is mainly because the dispersion of measured values in technical processes can be well described by such a distribution. It is given by

$$p_G(x; \mu, \sigma) = \frac{1}{\sqrt{2\pi\sigma^2}} \exp\left[-\frac{(x - \mu)^2}{2\sigma^2}\right] \tag{2.40}$$

and is described by the parameters μ and σ. μ indicates the position of the maximum of the distribution and σ the width. A helpful specification for the Gaussian function is, in addition to the variance, also the full width at half maximum (English: *full width at half maximum, FWHM*),

$$\text{FWHM}(\sigma) = 2\sigma\sqrt{2\log 2}, \tag{2.41}$$

where we denote the natural logarithm with the abbreviation "log". The skewness of the Gaussian distribution is $\nu = 0$ and the kurtosis is $\kappa = 3$, i.e., both higher moments are explicitly not dependent on the distribution parameters. You can easily verify this by inserting the function into the definition of the moments.

Poisson distribution

When events occur rarely, such as radioactive decay, this process is often described with a Poisson distribution. This distribution can be imagined as a clustering curve of a temporal distance. If an event occurs at time $t = 0$, then an immediate subsequent event is not very likely, so the density is low near zero. However, the probability increases with increasing time and also decreases after a maximum. The Poisson distribution is discrete and defined as

$$p_P(x; \lambda) = \frac{\lambda^x}{x!}\exp(-\lambda). \tag{2.42}$$

Its expected value and its variance are equal, $\langle x \rangle = \lambda = \langle (x - \langle x \rangle)^2 \rangle$ and both are uniquely determined by λ. Its skewness is $\nu = 1/\sqrt{\lambda}$ and its kurtosis is $\kappa = 3 - \frac{1}{\lambda}$.

Pearson-IV-Distribution

The Pearson-IV distribution is an example of the function family by K. Pearson [11]. It allows the adjustment of four location parameters μ, σ, ν and κ, which ultimately adjust the symmetry, skewness and kurtosis,

$$p_{\text{Pearson-IV}}(x) = \left[1 + \left(\frac{x - \mu}{\sqrt{2\sigma^2}}\right)^2\right]^{-\nu} \exp\left\{-\kappa\tan^{-1}\left(\frac{x - \mu}{\sqrt{2\sigma^2}}\right)\right\}. \tag{2.43}$$

This special Pearson distribution is used in the description of wave phenomena, in the collection of charge in capacitors, but also in asymmetric process corridors (which we will discuss in more detail later in Chap. 6, Sect. 6.5) with pronounced tailing to one side.

Weibull Distribution/Rosin-Rammler Distribution

Our last example of a probability density is the Weibull distribution. It represents a whole family of curves and can model left- and right-skewed as well as symmetric distributions. Its density is given by

$$p_W(x; \lambda, k) = \begin{cases} \lambda k(\lambda x)^{k-1}\exp\left[-(\lambda x)^k\right] & \text{für } x \geq 0 \\ 0 & \text{sonst,} \end{cases} \tag{2.44}$$

The Weibull distribution is of great importance, especially for the failure of technical components. It helps us to statistically express quality data such as the mean time to failure (MTF) or the mean time between failures (MTBF). Its expected value is

$$\langle x \rangle = \frac{1}{\lambda} \Gamma \left(\frac{k+1}{k} \right) \tag{2.45}$$

and its variance

$$\langle (x - \langle x \rangle)^2 \rangle = \frac{1}{\lambda^2} \left[\Gamma \left(\frac{k+2}{k} \right) + \Gamma^2 \left(\frac{k+1}{k} + \right) \right]. \tag{2.46}$$

. Here, we have used the gamma function as a shorthand, for which

$$\Gamma(n+1) = \int_0^\infty x^n e^{-x} dx \tag{2.47}$$

applies.

2.3.7 Matrix Representation of Multiple Measurement Series

In many cases, there will be multiple instances $j \in \mathbb{N}$ of measurements that lead to a set of $N \in \mathbb{N}$ different vectors x_j that can be arranged in a matrix,

$$X = \begin{pmatrix} x_{0,0} & x_{0,1} & \cdots & x_{0,M-1} \\ x_{1,0} & x_{1,1} & \cdots & x_{1,M-1} \\ \cdots & \cdots & \cdots & \cdots \\ x_{N-1,0} & x_{N-1,1} & \cdots & x_{N-1,M-1} \end{pmatrix}. \tag{2.48}$$

In summary, these matrices represent N measurements of the (same) variable x, with M data points for each measurement. Sometimes this matrix X is also referred to as a **data matrix**. Especially for the training of supervised and unsupervised machine learning algorithms, data in the form of such a matrix X for the representation of **training** and **test data sets** is essential.

A practical step often involves transposing the matrix X to obtain the appropriate form for further processing. This can be done in Python in various ways, of which we will briefly introduce only two common methods:

Listing 2.4 Example for a matrix in Python

```python
import numpy as np
X=[[1,2,3],[4,5,6,],[7,8,9]]
print(X)
print(np.transpose(X))
```

Listing 2.4 shows the transposition of a Matrix in Python using Numpy. The transposition works similarly with the Pandas library, as demonstrated in Listing 2.5.

Listing 2.5 Example for matrix using the library Pandas

```
import pandas as pd
X=[[1,2,3],[4,5,6,],[7,8,9]]

X=pd.DataFrame(X)
print(X)
print(X.T)
```

2.3.8 Covariance and Covariance Matrix

While the previous properties, expected value and variance, only refer to a single data series x, we now want to develop tools for comparing data. The simplest way to compare two functions is multiplication. If both functions increase or decrease equally, the product is positive. If both functions are opposite, the product is negative.

For two vectors with data x and y we define the **covariance** as

$$\text{Cov}(x,y) = \langle (x - \langle x \rangle)(y - \langle y \rangle) \rangle. \tag{2.49}$$

The covariance is a measure of the similarity of two data trends.

We perform a multiplication between both signals here to see where the signals overlap strongly and whether they are similar. For a measurement matrix X we can thus also set up a covariance matrix,

$$C = \text{Cov}(X) = \begin{pmatrix} \text{Var}(x_0) & \text{Cov}(x_0, x_1) & \ldots & \text{Cov}(x_0, x_{N-1}) \\ \text{Cov}(x_1, x_0) & \text{Var}(x_1) & \ldots & \text{Cov}(x_1, x_{N-1}) \\ \ldots & \ldots & \ldots & \ldots \\ \text{Cov}(x_{N-1}, x_0) & \text{Cov}(x_{N-1}, x_1) & \ldots & \text{Var}(x_{N-1}) \end{pmatrix},$$

where the variances represent the main diagonal and the subsequent covariances are symmetrically arranged in the upper right and lower left corners of the matrix. Please note that the covariance increases with increasing agreement of x and y. It is therefore a tool to find out how well two data series match. If x and y run uniformly, it is positive. It becomes negative if y is directed against the course of x and ultimately 0 if there is no connection at all.

The Pandas toolbox allows us to quickly evaluate the statistical key features of our data. Listing 2.6 shows the application of the "describe" command. It is applied to the transposed variant of the matrix X,

Listing 2.6 Extract statistical information using Pandas

```
1   import pandas as pd
2
3   pdf = pd.DataFrame(X)
4   pdf.T.describe()
```

which finally returns a table with properties:

Listing 2.7 Results of Listing 2.6

```
1   count 50.000000 50.000000 50.000000 ...
2   mean 326.016998 314.661663 328.752005 ...
3   std 236.253411 256.087978 240.108021 ...
4   min 73.663312 51.154939 88.234382 ...
5   25% 147.841750 111.452276 135.089759 ...
6   50% 241.469434 223.321377 247.337707 ...
7   75% 451.525597 451.451403 458.632592 ...
8   max 880.165305 984.095894 920.493881 ...
```

2.3.9   Correlation and Correlation Matrix

Investigation for Similarity

The covariance is not standardized. It still depends on the amplitudes of the variables x and y, and this can distort the view of the underlying relationships. Therefore, we move to a quantity that is standardized: If we normalize the covariance with the respective standard deviation of the data series, we arrive at the correlation:

> The **correlation** of two data vectors x and y is defined as
>
> $$\text{Cor}(x,y) = \frac{\text{Cov}(x,y)}{\sigma_x \sigma_y} = \frac{\text{Cov}(x,y)}{\sqrt{\text{Var}(x)\text{Var}(y)}}$$
>
> and indicates how similar the directional trends of both vectors are.

The correlation no longer depends on the amplitudes, only on the course of x and y. Thus, only the relationship between both data series is captured. Similar to the covariance matrix, the correlation matrix is given by,

$$\text{Cor}(X) = \begin{pmatrix} 1 & \text{Cor}(x_0, x_1) & \dots & \text{Cor}(x_0, x_{N-1}) \\ \text{Cor}(x_1, x_0) & 1 & \dots & \text{Cor}(x_1, x_{N-1}) \\ \dots & \dots & \dots & \dots \\ \text{Cor}(x_{N-1}, x_0) & \text{Cor}(x_{N-1}, x_1) & \dots & 1 \end{pmatrix}.$$

It also reflects that each data series is fully correlated with itself. Therefore, there is a 1 on the main diagonal of this matrix.

Implementation of the correlation matrix

To explain this with an example, we have generated four different functions in Listing 2.8 and correlated them with each other. We expect that the functions $x1$ and $x3$ are anticorrelated. We expect a fundamentally positive correlation between $x4$ and $x3$.

Listing 2.8 Calculating a correlation matrix using Numpy

```
import numpy as np

t = np.arange(0,4*np.pi, 0.1)
x1 = np.sin(t)
x2 = np.cos(t)
x3 = -np.sin(t)
x4 = t**2
C = np.corrcoef([x1, x2, x3,x4])
```

Feel free to take the opportunity to output C from this listing. We would like to visually represent the correlation matrix at this point and use Listing 2.9 for this.

Listing 2.9 Visualizing a correlation matrix using Matplotlib

```
import matplotlib.pyplot as plt

plt.imshow(C, aspect='auto')
plt.xlabel('Index')
plt.xticks(range(0,4), ['x1','x2', 'x3', 'x4'])
plt.yticks(range(0,4), ['x1','x2', 'x3', 'x4'])
plt.colorbar()
```

Fig. 2.6 shows the correlation matrix C for our example.

Correlation analysis as an exploratory tool

The correlation matrix represents a versatile tool of exploratory analysis. The representation in Fig. 2.6 provides a good first overview of the relationships. As mentioned in Chap. 1, a high correlation indicates that the variables may be related, but they do not necessarily have a causal relationship with each other.

Summary

In this chapter, we have revisited the basics of set theory, stochastic, and statistics. These disciplines provide us with tools that are essential for understanding machine learning. Thus, the division of data into a training set and a test set represents an important step for learning models.

Conditional probability generally serves as the basis for decision-making processes and will later become a fundamental tool for creating decision trees. Advanced methods such as mixture-density networks use probability distributions in the internal structure of the network itself and allow, through this step, the prediction of their own uncertainty.

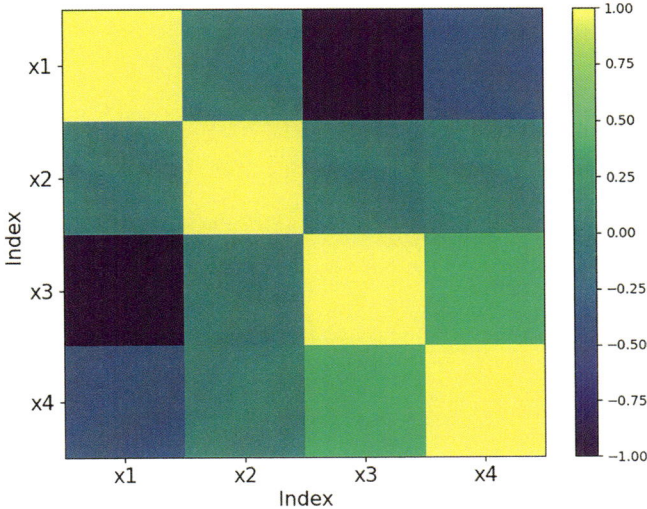

Fig. 2.6 Correlation matrix from the example code 2.8

The statistical properties of the data, above all their moments, are suitable for reducing information to a few data points. In many cases of physically-informed methods, they are an additional key piece of information that is added to the learning process.

The terms covariance and correlation will continuously accompany us. For instance, the reduction of the geometric covariance of cluster points plays a central role in the K-means method. As a means of exploratory analysis, correlation is capable of providing us with valuable preliminary information about our data. In real, practical work situations, it is always recommended to start with such a correlation analysis to get a first impression of the data.

Tasks

2.1 Determine the moments of the Poisson distribution by starting with equation (2.36).

2.2 Determine the accumulated probability distribution F for the dice process and represent it graphically.

2.3 Given are the synchronous event lists $L_1 = [1, 1, 0, 1, 0, 1]$ and $L_2 = [1, 0, 0, 1, 0, 0]$, where synchronous means that the first element of L_1 and the first element of L_2 occur simultaneously, etc. Determine $P(L_1 = 1 | L_2 = 1)$ and $P(L_2 = 1 | L_1 = 1)$.

2.4 Consider the lists $L_1 = [0.9, 0.5, 0, 0.7, 0, 0.4]$ and $L_2 = [0.9, 0, 0, 0.8, 0, 0]$ and repeat their calculation of $P(L_1 = 1 | L_2 = 1)$ and $P(L_2 = 1 | L_1 = 1)$. What has changed?

2.5 Show that the Gaussian distribution has the skewness $v = 0$ and the kurtosis $\kappa = 3$. Start with the definition of these moments in (2.36).

2.6 Represent the Weibull distribution for $\lambda = 1$ and $k = 2$ using Matplotlib in Python.

References

1. H.-D. Ebbinghaus, *Einführung in die Mengenlehre*. Springer Spektrum Berlin, Heidelberg, 2021.
2. L. Fahrmeir, R. Künstler, I. Pigeot, and G. Tutz, *Statistik*. Springer, 1997.
3. B. Hall and M. Kuster, „Representing quantities and units in digital systems," *Measurement: Sensors*, vol. 23, p. 100387, 2022. [Online]. Available: https://www.sciencedirect.com/science/article/pii/S2665917422000216
4. C. Henke, E. Jacobs, N. Teofilov, P. Henke, and M. J. Neuer, „Sensor fusion of spectroscopic data and gyroscopic accelerations for a direction indication in handheld radiation detection instruments," in *Proceedings of the IEEE Nuclear Science Symposium and Medical Imaging Conference, Strasbourg*, 2016.
5. S. C. Hora, „Aleatory and epistemic uncertainty in probability elicitation with an example from hazardous waste management," *Reliability Engineering and System Safety 54(2–3), 217–223*, 2010.
6. E. Hüllermeier and W. Waegeman, „Aleatoric and epistemic uncertainty in machine learning: an introduction to concepts and methods," *Machine Learning 110, 457–506*, 2021.
7. D. e. a. Hutzschenreuter, „SmartCom Digital System of Units (D-SI) Guide for the use of the metadata-format used in metrology for the easy-to-use, safe, harmonised and unambiguous digital transfer of metrological data," Nov. 2019, The development of the uniform metadata format for the exchange of measurement data in ICT applications is part of the research project EMPIR 17IND02 (Title: SmartCom). This brochure is the result of Deliverable 1. The project was funded by the EMPIR programme co-financed by the participating countries and by the European Union's Horizon 2020 research and innovation programme. [Online]. Available: https://doi.org/10.5281/zenodo.3522631
8. S. J. Julier, T. Bailey, and J. K. Uhlmann, „Using exponential mixture models for suboptimal distributed data fusion," *IEEE Nonlinear Statistical Signal Processing Workshop*, pp. 160–163, 2006.
9. A. Kolmogorov, „Interpolation and extrapolation of stationary random sequences," *Izvestiya AN SSSR*, vol. 5, p. 314, 1941.
10. S. Maskell, „A bayesian approach to fusing uncertain, imprecise and conflicting information," *Information Fusion*, vol. 9, no. 2, pp. 259–277, 4 2008.
11. K. Pearson, „Contributions to the mathematical theorie of evolution," *Proc. Roy. Soc. London*, vol. 54, pp. 329–333, 1893.
12. K. Pearson, „Contributions to the mathematical theorie of evolution. ii. skew variation in homogeneous material," *Phil. Trans. R. Soc. Lond.*, vol. A, no. 186, pp. 343–414, 1895.
13. K. Pearson, „Mathematical contributions to the theory of evolution. x. supplement to a memoir on skew variation," *Phil. Trans. R. Soc. Lond.*, vol. 197, pp. 443–459, 1901.
14. C. E. Shannon, „A mathematical theory of communication," *The Bell System Technical Journal*, vol. 27, pp. 379–423, 623–656, 7 1948.

15. H. Toutenburg, M. Schomaker, and C. Heumann, *Induktive Statistik*, ser. Springer-Lehrbuch. Springer, 2008. [Online]. Available: https://books.google.de/books?id=Gof0oZmxy2kC

16. E. Vanden-Eijnden and R. Balescu, „Statistical description and transport in stochastic magnetic fields," *Phys. Plasmas*, vol. 3, p. 874, 1996.

17. N. VanKampen, „Stochastic processes in physics and chemistry," *Phys. Fluids*, vol. 19, p. 11, 1996.

18. M. Vetterli, P. Marziliano, and T. Blu, „Sampling signals with finite rate of innovation," *IEEE Trans. Sign. Proc.*, vol. 50, no. 6, pp. 1417–1428, 2002.

Chapter 3
Data Preprocessing

Keywords Exploratory data analysis · Filtering of data · Data transformations · Process parameters

Data preprocessing is an important element in the process chain of machine learning methods. In this chapter, we discuss various methods to achieve optimal preparation for individual problem cases. This includes normalization, triggering, filtering, and also mathematical transformations such as the Fast Fourier Transformation (FFT) or the continuous wavelet transformation. The chapter also shows how to extract and use probability distributions from data.

Fig. 3.1 shows you how the contents of Chap. 3 influence the subsequent chapters. Preprocessing is significant for all learning methods. An important aspect is the feedback from Chap. 6. Here, the physically informed method can request specialized preprocessing steps. For this reason, we focus our selection of topics for Chap. 3 on steps that require some prior knowledge. The Fourier transformation and the extraction of characteristic sizes are examples of this. Of course, the learning methods presented in Chap. 4 and 5 can also be used in a larger context as preprocessing.

Which steps data have gone through in a data processing chain is also relevant for explainability. It is important to be able to explain why a certain type of preprocessing was chosen. While there are learning methods that could learn any transformation without further ado, we would then lose the traceability that we want to achieve for explainable methods. The link between preprocessing, learning methods, and ultimately a digital capture of the meaning of these steps is ultimately a central element in achieving the explainability of the algorithm.

M. J. Neuer, *Machine Learning for Engineers*,
https://doi.org/10.1007/978-3-662-69995-9_3

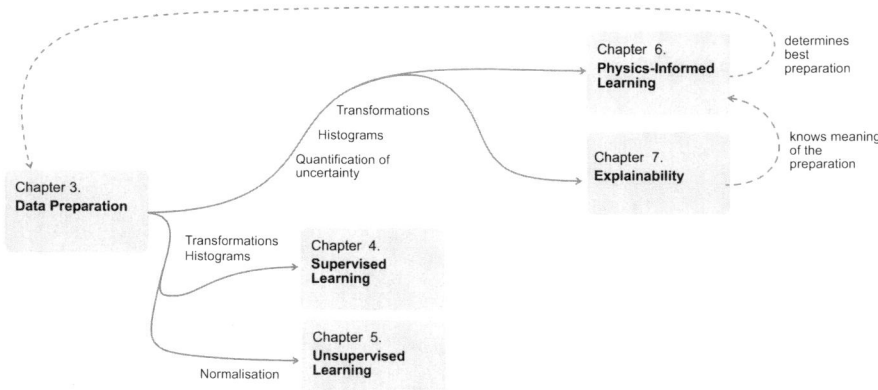

Fig. 3.1 Overview of the relationship of Chap. 3 with the following chapters

3.1 Goals of Preprocessing

Data preprocessing encompasses all necessary steps that generate a meaningful dataset from a set of raw data. Meaningful are all processes that maintain and, if possible, increase the interpretability of the data.

Data preprocessing includes the following steps, which we want to discuss in more detail in this chapter:

- **Data cleaning.** Here, the dataset is examined for problems and prepared. In practice, these problems occur more frequently and this step is often labor-intensive. We can distinguish the following sub-points:
  - Find and replace Not-a-Number entries (NaNs),
  - Find and replace missing data, possibly supplement with interpolation,
  - Detect duplicates and possibly remove,
  - Check and correct lengths of time series.
- **Normalization.** Normalization rescales the data to a different numerical range. It defines the scale on which we want to work. Especially when comparing variables, this intervention is helpful. For machine learning methods, it can play a major role whether the data is passed in the correct scaling or not.
- **Filter.** In signal theory, filters play an important role. Machine learning methods can also benefit from this filter, as they help to smooth data (moving average) or to enhance certain properties of data (e.g., through the first and second derivative).
- **Triggering.** Especially with long time series with recurring signal sequences, triggering is often necessary. The repetitive part is then cut out, can be overlaid and examined for outliers.
- **Application of a function.** Sometimes it can be helpful to transform the data variables via a function. There are many variants here; the formation of the

logarithm, to better recognize the dynamics in exponents from exponential dependencies, is an example of this.

- **Transformation.** Transformations are comparable to normalization, but they also change the qualitative course of the variables or transfer them into a perspective that opens up a better, perhaps even low-dimensional, view. The goal of every transformation is to improve the later evaluability.
- **Statistical and process-related characteristics.** Here we determine important points in the course of data. These can be extremes, averages or other statistical sizes, but also positions motivated by process understanding, which are important for a technical reason. Examples are breakthrough points in air filters, saturation points, reaching half-lives or turning points in motion sequences.
- **Application of further algorithms.** In principle, the difference between evaluation methods and preprocessing is so fluid that we can consider almost any algorithm as data preprocessing, as long as it can be used sensibly. Especially the unsupervised learning methods, of which we will get to know the principal component analysis, K-means clustering method and the autoencoder, are excellently suited for data preparation.

3.2 Data Cleaning

3.2.1 *Removal of Faulty Data Points*

A typical task of data preprocessing is data cleaning. Analysis algorithms cannot function properly with damaged or incomplete data. An example of this is the occurrence of Not-a-Number entries (NaNs). There are several reasons why the data contains NaNs: either because bad signals were already generated by the sensor or simply because the signals had a problem at some stage of processing.

In many raw data, you find such faulty values, missing values or wrong scalings. Where do these problem spots come from? The path that an entry in a database or in a file takes can be complex. Errors can occur during measurement, for example, when the measured value leaves the prescribed range. They can occur when saving the data, if the write operation was wrong. When working with databases, it may be that the query is not correct or an error occurred during the transfer of data over the network.

Always clean up at the source first
When confronted with data errors, you should find out where they come from. Which process was responsible for the errors? Often, data errors indicate another, deeper problem. Once you have identified this process, check whether the writing of NaNs or the causal error in general can be prevented.

Example: Weather Station
A weather station in your garden records the humidity and air pressure. It is powered by a solar cell. The data is transmitted to a computer via Bluetooth. The computer writes at the same rate, e.g. 1/s, and it writes NaNs as soon as the transmitter does not send a value. Whenever the voltage of the solar cell is insufficient, the transmission is interrupted. Therefore, you receive a series of data with NaNs, in which a meaningful value for their measurements always occurs when the voltage was sufficient. In this case, it would be helpful to first improve the voltage of the measuring device and optimize the transmission chain as much as possible. ◄

Example: QR Scanner
Due to a dirt spot on a QR scanner (camera), the QR code of products is no longer read correctly. From a certain point in time, there are erroneous data in your database. Here too, it makes sense to remedy the situation by physically cleaning the camera. ◄

These examples are intended to show you how important the quality of the measurement chain is for the quality of the data. Do not try to correct errors in a measurement chain by repairing digital data. This would mean addressing the error in the wrong place. In industrial companies, agreeing on fixed maintenance and cleaning measures can also help to improve data quality.

Cleaning of raw data
Correcting problem areas directly at the data source is therefore, according to our previous considerations, always the best way to ensure data quality. However, in practice we often find ourselves confronted with historical data sets whose purely sensory information we can no longer improve afterwards. Or we technically have no real access to repair the measurement chain, e.g. when sensors are built in inaccessible. Ultimately, and this is usually the most important argument, the necessary correction of the sensor or the measurement chain may simply be uneconomical. In such cases, we must perform an adequate cleaning of the data on our side.

In Listing 3.1, a table is created and then re-indexed with Pandas so that it contains more rows than there are data. The only reason for this re-indexing is the creation of rows with NaNs.

Listing 3.1 Replacing NaNs in data

```
import numpy as np
import pandas as pd

frame=pd.DataFrame(np.random.randn(4,3),
                    index=[1,2,4,7],columns=['A','B','C'])

# create NaNs "artificially" by expanding the number of
# rows of the data frame
frameWithNaNs = frame.reindex([1,2,3,4,5,6,7])
replacedNaNs = frameWithNaNs.replace({NaN:0.0})

print(frameWithNaNs)
print(replacedNaNs)
```

If you run the above code from Listing 3.1, you get the output 3.2. Rows 1 to 8 contain the original matrix with NaN values. Rows 9 to 16 show the same matrix, but with the replacement NaN = 0.0.

Listing 3.2 Output of Listing 3.1

```
          A          B          C
1  1.015877  -0.194974  -0.777067
2  0.199423  -1.477063   0.679932
3       NaN        NaN        NaN
4  2.039111   0.908888   0.695052
5       NaN        NaN        NaN
6       NaN        NaN        NaN
7 -1.253851   0.255705  -0.569040
          A          B          C
1  1.015877  -0.194974  -0.777067
2  0.199423  -1.477063   0.679932
3  0.000000   0.000000   0.000000
4  2.039111   0.908888   0.695052
5  0.000000   0.000000   0.000000
6  0.000000   0.000000   0.000000
7 -1.253851   0.255705  -0.569040
```

Such replacements are of course not only possible with the toolbox Pandas. It is used here only as an illustrative example. You should also handle the replacement of data very consciously. The above code increases the number of occurrences of 0.0. If this number is relevant for your evaluation, e.g. if you are conducting a statistical evaluation, then such a replacement can also lead to errors.

The choice of the right replacement depends on the specific situation. Often you want to insert a numerical marker in the data to trace where the NaNs occurred. Suitable for this are numerical values that do not occur in the regular value range of the respective sensor. For a temperature sensor that measures between $-50\,°C$ and $100\,°C$, $-1000\,°C$ would be a suitable marker.

3.2.2 Missing Data

Should data be incomplete, e.g., because there is an incorrect value at an index position or a NaN replacement has been made, the data set can be completed by direct interpolation. Let's assume that the defective data position is at x_i. If both x_{i-1} and x_{i+1} contain trustworthy values, we can use

$$x_i = \frac{x_{i-1} + x_{i+1}}{2} \tag{3.1}$$

to implement a manual interpolation. The mean value represents the most probable value at exactly this position.

3.2.3 Tracking and Eliminating Duplicates

With the following simple listing, we can effectively search for duplicates in our data frame. To practice this, we have created a very simple set of vectors (without any meaning) to see how Pandas enables the check for duplicates. The listing 3.3 shows an example of searching for duplicates.

Listing 3.3 Searching for duplicates using Pandas

```
import numpy as np
import pandas as pd

x0 = [1,2,3,4,5,6,7,8,9]
x1 = [1,3,1,4,4,1,2,5,1]
x2 = [1,3,1,4,4,3,2,5,5]
x3 = [1,3,3,3,1,2,4,5,1]

frame = pd.DataFrame([x0,x1,x2,x1,x3])
frame.duplicated()
```

As a result of the search in the Pandas DataFrame, we get in listing 3.4 a vector with True/False entries, which shows us whether and where the duplicate can be found.

Listing 3.4 Result of Listing 3.3

```
0       False
1       False
2       False
3        True
4       False
dtype: bool
```

Due to the constructed example, it turns out that the vector x1 occurs exactly twice and the duplicate is at the penultimate position. In fact, we have inserted x_1 twice into the data frame, at position $i = 1$ and at position $i = 3$.

3.3 Normalization

3.3.1 Reasons for Normalizing Data

In Chap. 1 we discussed scales and scale levels discussed. They help us understand the nature of a data series. The data type also determines the possible mathematical operations that we can perform with it. In the example of the different relative temperature differences in the Celsius and Kelvin scales from 1.2.7 we could already see how important the measurement range is for our own perception. This effect depends on the scaling of our data and we can influence this with the help of suitable rescaling.

Example: Air Pressure
Let's take another example to help clarify this. Depending on the altitude, the air pressure varies between 1070 hPa and 800 hPa (on high mountains). So, at 800 hPa we would speak of very low pressures and at 1070 hPa of very high pressures. If our data refers to tire pressures, which are rather in the range of 200 kPa, we have a unit jump from hPa to kPa. Our previous assessment of low and high would be obsolete. ◀

3.3.2 Types of Normalizations

Often, the scaling of our data needs to be adjusted between different datasets. Data is normalized to become comparable. The following normalizations help with this:

- **Normalization to the maximum of the data vector.** The data is divided by its maximum,

$$x = \frac{x}{\max(|x_i|)},\tag{3.2}$$

which results in a scaling of the individual data points within the interval $[-1, 1]$.

- **Normalization to the sum of the data vector.** The data is divided by its total sum (magnitude of the data vector),

$$x = \frac{x}{\sum_i x_i},$$ (3.3)

which results in each data vector having the same area after normalization.
- **Min-Max Normalization.** Sometimes it is necessary to only consider positive data points for an evaluation. Then our data can be mapped to the values [0,1] by first subtracting the minimum value and then scaling to the largest distance,

$$x = \frac{x - 1\min(x)}{\max(x) - \min(x)}.$$ (3.4)

- **Subtraction of the mean.** Another form of normalization ensures that our resulting data has a new average of $\langle x \rangle = 0$. Here, the mean is first subtracted and then divided by the distance between maximum and minimum,

$$x = \frac{x - 1\langle x \rangle}{\max(x) - \min(x)}.$$ (3.5)

Normalization is a common source of error and should therefore be carried out with care. One can unintentionally change the meaning of data and in the worst case reduce or lose the information content of variables. However, normalization is necessary for the later field of physically informed machine learning. It helps to focus on relevant aspects of the data. Important areas of measurements are specifically highlighted.

3.3.3   Application of Normalization

To emphasize the importance of normalization, we would like to consider a simple example that will accompany us through the rest of the book.

Example: Motor Current Anomaly
You are looking at a composite of motors. The motors are identical and provide you with current characteristics. These characteristics contain information about whether the process ran correctly or not. The latter is comparable to an impending defect of a motor or problems with their application. However, an error occurred during the recording of the data and the automatic detection of the measuring range measured the current in milliamperes in some cases and in A in other cases. ◀

We first load the data and look at it as we have outlined in Listing 3.5.

Listing 3.5 Loading an example data file and plotting the contained data

```
import matplotlib.pyplot as plt
import numpy as np
import pickle

data = pickle.load(open('EX03EngineExt.pickle', 'rb'))
for i in range(0,500):
    plt.plot(data['X'][i], color='k', alpha=0.25)

plt.xlabel('t', fontsize=18)
plt.ylabel('X', fontsize=18)
plt.tick_params(labelsize=18)
```

The result is shown in Fig. 3.2: some current profiles are clearly visible, while another group appears as a horizontal line in the diagram. The latter are exactly the data that were measured in the ampere range, so they have much smaller values than the measurements in mA. We have not yet noted a unit on the axis, as we are only working purely exploratively here and are initially only looking at the data.

We want to normalize the data so that all characteristics have the same order of magnitude. Therefore, we change our existing code by inserting Listing 3.6 from line 5. This shows an extremely simple, but effective, manual triggering—a pre-processing step that we will revisit in Sect. 3.5. Whenever the maximum of a curve is large enough, we let it apply, in all other cases we scale the curve by a factor of 1000 to switch from the A to the mA scale.

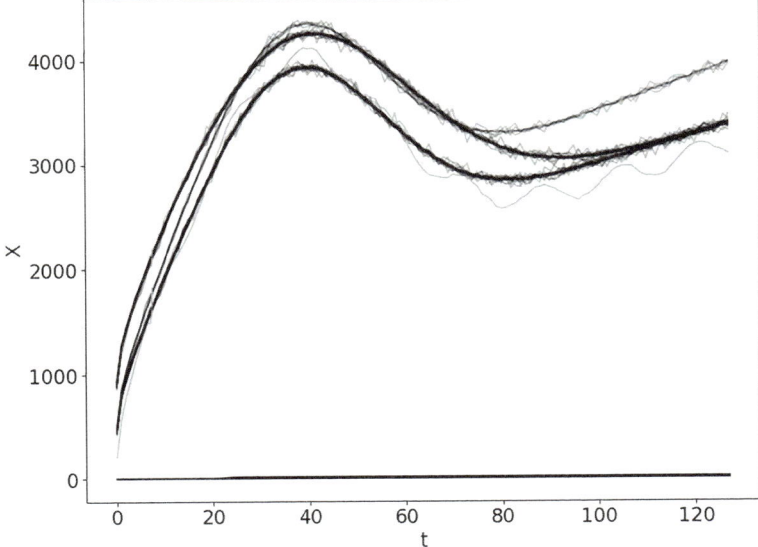

Fig. 3.2 Result of Listing 3.5

Listing 3.6 Applying a simple trigger

```
data = pickle.load(open('EX03EngineExt.pickle', 'rb'))
newData = []
for i in range(0,500):
    if np.max(data['X'][i]) > 1000:
        newData.append(data['X'][i])
    else:
        newData.append(1000*data['X'][i])

    plt.plot(newData[-1], alpha=0.2)
```

The example is intended to show you how easy it can be to fix the error of the measuring ranges. However, it is not always clear that such an error exists. We want to briefly present the other normalization variants.

Listing 3.7 shows the procedure to normalize to the sum of the data vector. Here we go through the same for-loop as for the correction of the units.

Listing 3.7 Sum-Normalization

```
sumNormalizedData = []
for i in range(0,500):
    normalised = newData[i] / np.sum(newData[i])
    sumNormalizedData.append(normalised)
    plt.plot(sumNormalizedData[-1], alpha=0.2)
```

If it is important to normalize to the maximum, Listing 3.8 shows how the reference to the maximum per data vector is calculated with minimal changes.

Listing 3.8 Maximum Normalization

```
maxNormalizedData = []
for i in range(0,500):
    normalised = newData[i] / np.max(newData[i])
    maxNormalizedData.append(normalised)
    plt.plot(maxNormalizedData[-1], alpha=0.2)
```

Finally, there is a normalization where we first subtract the mean and then normalize to the maximum. This is shown in Listing 3.9. But let's not confuse this variant with a normalization to the resulting maximum after subtraction.

Listing 3.9 Subtraction of mean

```
meanMaxNormalizedData = []
for i in range(0,500):
    normalised = (newData[i] - np.mean(newData[i])) / np.max(
        newData[i])
    meanMaxNormalizedData.append(normalised)
    plt.plot(meanMaxNormalizedData[-1], alpha=0.2)
```

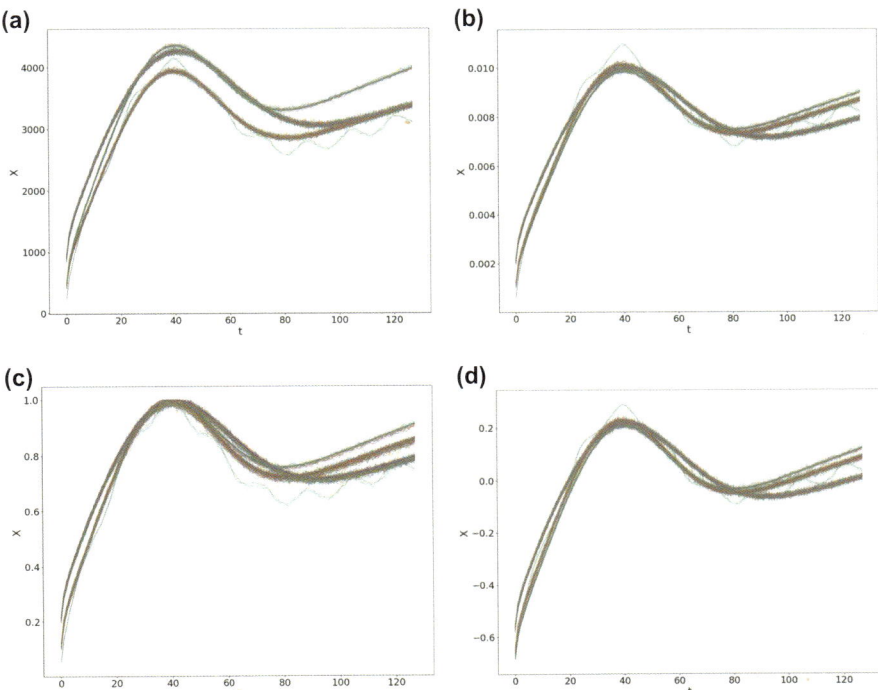

Fig. 3.3 Impact of different normalizations on the motor current dataset. (a) No additional normalization, (b) Normalization to the sum (3.3), (c) Normalization to the maximum (3.2) and (d) Subtraction of the mean and normalization to the maximum

In Fig. 3.3 the normalization variants shown here are presented. Each normalization has a different effect. With some normalizations, you can amplify or attenuate effects. Whether this is useful or whether the normalization is disadvantageous for your data analysis depends on the specific example.

3.4 Filtering of Data

3.4.1 Moving Average

One of the most important transformation steps in data processing and data preparation is smoothing. Data that is too noisy must be smoothed so that subsequent analysis algorithms do not confuse the inherent noise with the actual information content. The classic filter for denoising a signal is the moving average. In Listing 3.10, an example implementation of this filter is shown. We examine the effect of this filter using a simple example:

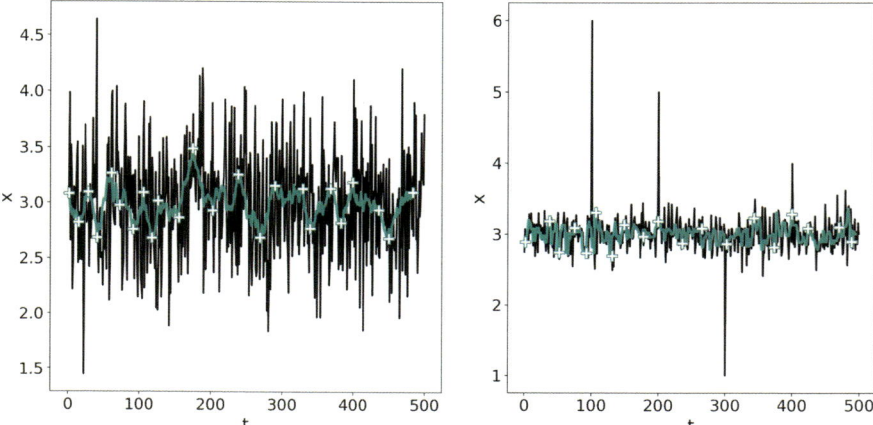

Fig. 3.4 Effect of the moving average (left) and effect of the moving median filter (right). The black line shows the raw data, the green (gray) line with white crosses shows the smoothed data

Listing 3.10 Moving average filtering

```
import numpy as np

x = 3+0.5*np.random.randn(500)
filtered_x = []
window = 15

for i in range(0,len(x)-window):
    filtered_x.append(np.mean(x[i:i+window]))
```

The variable x is a combination of 3.0 and a random number between -0.5 and 0.5. x is to be transformed into the filtered variable `filtered_x`. For this, we use the for-loop in lines 7 and 8. Here, the mean of `np.mean(x[i:i+15])` is first calculated and then this value is stored in `filtered_x`.

This specific implementation calculates the mean always forward. The for-loop must take this into account and therefore stops in this case 15 values before the end. With the code in Listing 3.11, the effect of the filter can be observed. Fig. 3.4 shows on the left side the noisy signal in black and the moving filter value in green. The noise is effectively smoothed.

Listing 3.11 Plotting the moving average

```
import matplotlib.pyplot as plt
plt.plot(x, 'k', linewidth=2.0)
plt.plot(filtered_x,'-',color=[0.1,0.65,0.6],linewidth=3.0,
    alpha=0.9)
plt.xticks(fontsize=18)
plt.yticks(fontsize=18)
plt.xlabel('t', fontsize=20)
plt.ylabel('x', fontsize=20)
```

The filter is described by a parameter, which is the window size, in Listing 3.10 named by the variable `window`. The larger the window, the stronger the smoothing effect, as the average is taken over a larger number of points.

3.4.2 Convolution

The sliding average filter shown here can be implemented more simply by using the convolution operation. This is exemplified in program example 3.12, with a smoothing window of sample length 15. For this, we use the function `np.ones(N)`, which generates an array with N ones for us. This shaping function, in our case a series of ones, is also called a convolution kernel.

Listing 3.12 Moving Median Filter

```
x = 3+0.5*np.random.randn(500)
window=15
filtered_x = np.convolve(x,
        np.array(np.ones(window)/window,
        mode='valid')
```

Let's take a closer look at this property for the moment. The convolution also helps us to perform other processing stages of our data quickly.

Convolution Formally, the convolution operation ∘ for two continuous functions $x(t)$ and $y(t)$ is defined by

$$(x \circ y)(\tau) = \int x(t)y(\tau - t)dt. \tag{3.6}$$

In the case of two data vectors x and y the discrete convolution operation is defined as

$$(x \circ y)_i = \sum_k x_k y_{i-k}. \tag{3.7}$$

The moving average filter, as discussed in the previous section, can be expressed extremely compactly using

$$MA(x) = f \circ [1, 1, 1, \ldots, 1]/N, \tag{3.8}$$

which was also implemented in code example 3.12.

3.4.3 The Median Filter

The median $m(x)$ of a vector x with sample data points is defined as the value x, which separates the sample so that 50 % of all points lie below x and the other 50 % of all points lie above x. Formally, the median can be written as

$$m(x) = \begin{cases} x_{(n+1)/2}, & \text{if } n \text{ is odd,} \\[2ex] \frac{x_{n/2} + x_{(n/2)+1}}{2} & \text{if } n \text{ is even,} \end{cases} \tag{3.9}$$

where n is the length of the vector x. One of its most important properties is to remove extreme outliers from data sets. Consider the following example,

$$\mu([1, 1, 5, 1, 1]) = 2.6,$$
$$m([1, 1, 5, 1, 1]) = 1.0.$$

Given the peculiar vector $x = [1, 1, 5, 1, 1]$, the median m apparently completely ignores the influence of the highest number 5.

We can also use the median shown in 3.4.3 to filter the data series by applying it in the same way as the mean within a window on the data.

Listing 3.13 Gleitender Median-Filter

```
%pylab
import matplotlib
import numpy as np

x = 3+0.2*np.random.randn(500)
x[100]=6
x[200]=5
x[300]=1
x[400]=4
filtered_x = []

for i in range(0,len(x)-5):
    filtered_x.append(np.median(x[i:i+5]))
```

With the code in 3.11, the result of Listing 3.13 can also be displayed. In Fig. 3.4, the effect of the moving median filter is shown on the right side.

3.5 Triggering

3.5.1 Time series and repetitive data

There are technical processes that record their data in long time series. Often, however, the actually interesting time range is only a localized phenomenon. Large areas of the data series are therefore not relevant at all. In such situations, one would like to cut out the relevant sequences and examine them specifically. This process is called triggering. A trigger is a switch that engages under a certain condition and records the data along with it.

In Fig. 3.5, an example of such data is shown. The data for the example was synthetically produced, but is strongly based on real situations such as the rolling process, the switching on and off of circuits, or laser activations.

3.5.2 Implementation of the trigger

In Listing 3.14, we have implemented a trigger as an example. The trigger is tested on the dataset EXAMPLE02.pickle. Often, data recording tools already offer trigger functionality. The method shown here serves for understanding. The function is passed a time series x. In addition, the parameters threshold, i.e., a limit value for x, and windowLength, the length of the triggered area, are to be passed. Whenever x exceeds the limit value, a corresponding window is cut out. This general form of the trigger can be rewritten for any triggering factors or conditions. It is therefore extremely easy to adapt to other problems.

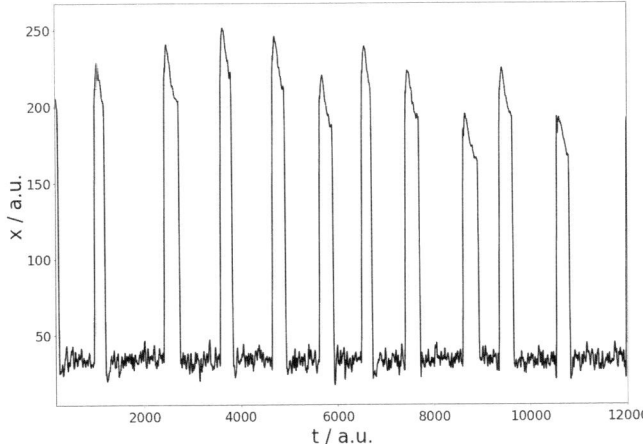

Fig. 3.5 Time series with repetitive process behavior

Listing 3.14 Triggering on a continuous signal

```
 1   x = pickle.load(open('EXAMPLE02.pickle','rb'))
 2
 3   def trigger(x, threshold, windowLength):
 4       triggerBuffer = []
 5       newTriggeredSection = []
 6       triggerStart = False
 7       isTriggerOn = False
 8       q=0
 9       for i in range(0,len(x)):
10
11           if x[i]>=1*threshold:
12               isTriggerOn = True
13               q=0
14
15           if isTriggerOn:
16
17               # -->
18               #if len(newTriggeredSection) < windowLength:
19                # newTriggeredSection.append(x[i-50])
20
21               # -->
22                newTriggeredSection.append(x[i-50])
23
24               q+=1
25
26           if q>= windowLength+50:
27               q = 0
28               isTriggerOn = False
29               newTriggeredSection = []
30               triggerBuffer.append(newTriggeredSection)
31
32       return triggerBuffer
33
34   data = trigger(x, threshold=70, windowLength=350)
```

You can also easily improve the implementation. It is often not necessary to iterate through the entire data series x. Whenever the trigger switches to True, you could directly jump to the end of the window length and continue iterating from there. Depending on what the specific problem looks like, there are therefore much more efficient trigger approaches.

The result of the triggering is shown in Fig. 3.6. Here, the data was also normalized to the maximum in order to better overlay the curves. You can already see from this type of representation whether individual curves show different behavior. A slight change in the code in 3.14, marked by the arrows in the comment, results in all windows being the same length.

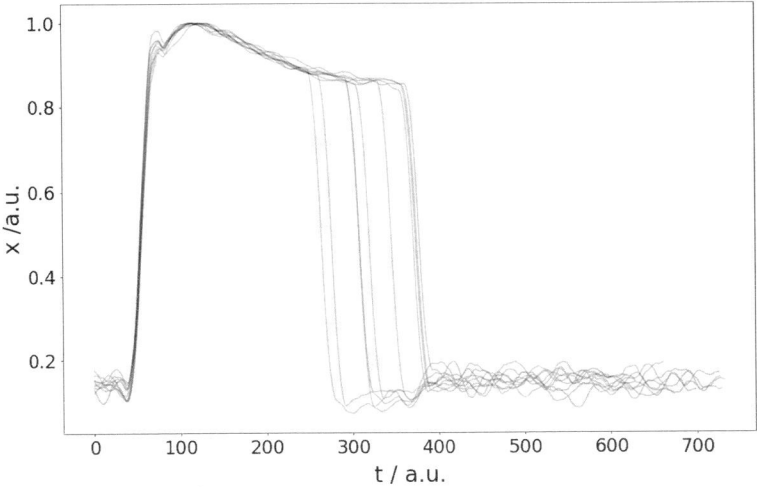

Fig. 3.6 Triggered data, each repetitive process step is represented by a curve

3.6 Transformations

3.6.1 *Differentiation of Data*

The derivative $f'(x)$ of a function $f(x)$ gives as is well known the slope at the point x. If a function is constant in x, the derivative is 0. In data, the derivative thus helps to eliminate constant parts and to amplify increases, decreases, any form of variation. The second derivative is a further stage of this procedure.

With the help of the convolution introduced in (3.7), the first derivative can be calculated over

$$\frac{df}{dx} = f \circ [-1, 1] \qquad (3.10)$$

If you apply the convolution kernel $[-1, 1]$ one more time, the second derivative results,

$$\frac{df}{dx} = f \circ [1, -2, 1]. \qquad (3.11)$$

If you are faced with particularly noisy data, it is advisable to combine the derivative with a smoothing average. Since deriving determines the slopes and noisy data practically consist of many slopes, you would otherwise amplify the noise. The combination of smoothing and derivation is

$$\text{Filter}(x) = f \circ [-1, -1, -1, -1, 1, 1, 1, 1]/4, \qquad (3.12)$$

where we have freely chosen the smoothing length 4. It must be adapted to the respective problem. Souza et al. show in [2] and how the application of such preprocessing can be systematically integrated into the process of data mining. Especially the differentiation is used in [5] to highlight the visibility of effects for machine learning methods.

The effect that a simple differentiation has on a data vector is shown in Fig. 3.7. The associated listing 3.15 shows how this example was constructed. We use the function diff(y,n) for the differentiation, where y represents the input data and n the degree of differentiation.

Listing 3.15 First and second derivative using Numpy

```
import matplotlib.pyplot as plt
import numpy as np

x = np.arange(0,100)
y1 = 0.5* np.exp(-(x-40)**2/25)
y2 = 0.2* np.exp(-(x-30)**2/25)
y3 = y1+y2+0.1*x

fig, (ax1, ax2, ax3) = plt.subplots(1,3,figsize=(16,9))
ax1.plot(x,y3,'k', linewidth=2.5)
ax2.plot(np.diff(y3),'k', linewidth=2.5)
ax3.plot(np.diff(y3,2),'k', linewidth=2.5)
```

In the left diagram of Fig. 3.7 you can hardly recognize the two maxima of the Gaussian function. The linear course of the curve dominates. The first derivative reduces this course. In the given example, the slope was 0.1 and the diagram in the middle of the figure shows the derivative, whose constant parts start at 0.1. A second derivative transforms this into a course around 0, because the constant part is eliminated by the further derivative. If these data were associated with a technical process, you would best highlight the effect shown here with derivatives.

3.6.2 Functional Transformation

Of course, you can apply any function to a data vector. This is always useful when you already suspect a certain trend,

$$x \mapsto \tilde{x} = \mathcal{F}(x; \pi), \tag{3.13}$$

where x stands for your original data series, $\mathcal{F}$ can be any transformation function and π captures the parameters of this function. The exponential function is a good example here. If you find exponential trends in your data, the logarithm is obviously a suitable transformation. This way, as when plotting on logarithmically scaled paper, you transform your data into a quasi-linear representation. In this case, no parameters would be necessary.

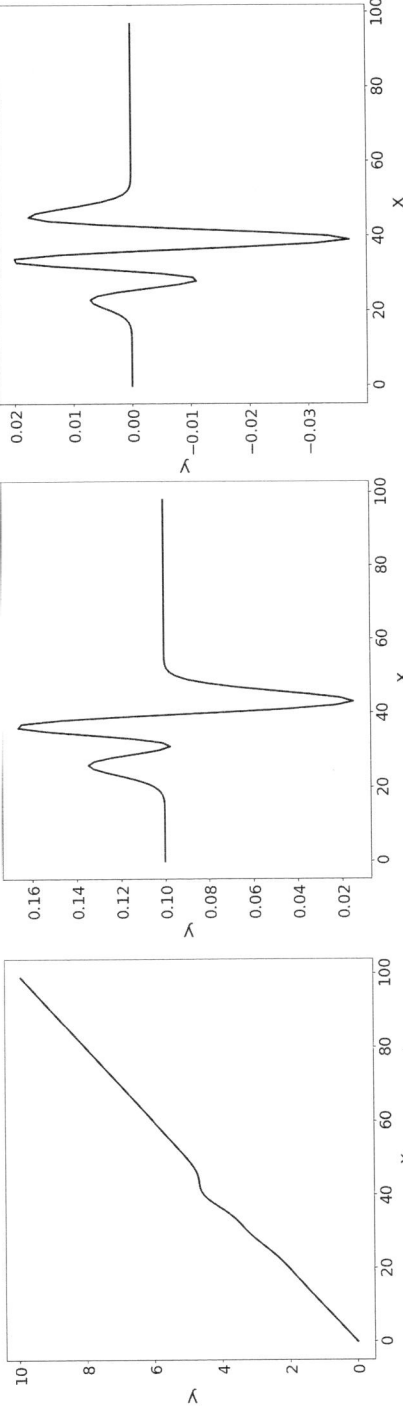

Fig. 3.7 Effect of the derivative on a data vector. Left: original course. Middle: first derivative. Right: second derivative

Applying a functional transformation is important for the later discussed phys-ically-informed learning methods and for the discussion of sensitivity analyses. Physically-informed learning uses such transformations to move from one input variable to an alternative variable. Sensitivity analysis can use this form of trans-formation in trained networks to check the analytical dependency between the result of a learning method and its input variables.

The following notes are important from a practical point of view for applying a direct function to your data:

- Do not increase complexity. For example, do not artificially create an oscilla-tion or another complex trend from the constant trend of your variables. This would contradict the purpose of preprocessing and above all only unnecessarily complicate your further processing steps.
- Avoid many parameters. Each parameter you introduce here acts as an additional adjustment screw in the learning methods. And since learning methods already bring many parameters with them, this greatly increases the depth of variation.
- If you try such a transformation, beware of new defective data after applying the function. If your dataset contains many zeros, divisions can lead to NaNs because you cannot divide by zero. Roots of negative values lead to complex results that cannot be used by every downstream method.

3.6.3 Fourier Transformation

Often we find recurring patterns in data, e.g. oscillations. They are characteristic of the signal trend. To make these patterns more visible, we can highlight special aspects from the signal through mathematical preprocessing.

The Fourier transformation calculates the frequencies and amplitudes of these signal components—it transforms from the space of temporal trends to the fre-quency space (also called Fourier space). For engineering sciences or for natural sciences in general, it is a known tool to describe oscillatory phenomena and to quickly switch between time and frequency spaces (Fig. 3.8).

The **Fourier Transform** $\mathcal{F}[x(t)](v)$ of a function $x(t)$ in the continuous case is

$$\mathcal{F}[x(t)](v) = \frac{1}{\sqrt{2\pi}} \int_{-\infty}^{\infty} x(t) \exp(-i2\pi vt)dt \qquad (3.14)$$

and in the discrete case, with discrete time steps Δt,

$$\mathcal{F}(v) = \frac{1}{\sqrt{2\pi}} \sum_{k=0}^{N-1} x(k\Delta t) \exp(-i2\pi vk\Delta t). \qquad (3.15)$$

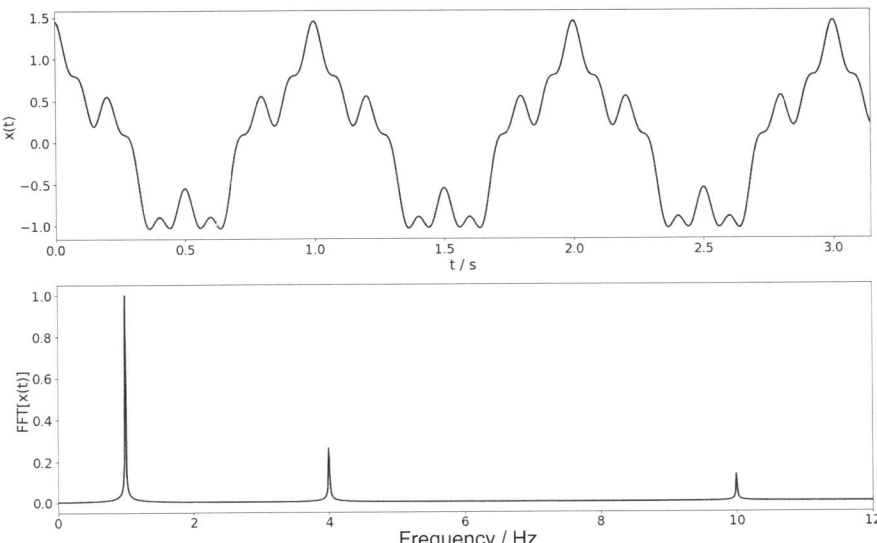

Fig. 3.8 Fourier transform from our code example with self-generated data

It develops the original course of the data into a series of oscillations $\exp(-i2\pi\nu k\Delta t)$. Here, i stands for imaginary numbers and is defined via $i^2 = -1$. Each oscillation is uniquely characterized by its frequency ν: They are localized in Fourier space, i.e., assigned to exactly one point on the frequency axis.

Since the calculation of the Fourier transform is time-consuming on a computer—the algorithm for this would be of the order (n^2) –, a method was developed which keeps the number of calculation steps for determining 3.15 small. This method is called Fast Fourier Transform (FFT). It divides the original time series into two new, smaller series using the individual time steps: the odd time steps go into one, the even time steps into the second new series. Both newly created time series are shorter than the original and can be transformed individually. Of course, the division process repeats recursively. The procedure thus created is called the Cooley-Tukey algorithm and is currently the most commonly used variant of the FFT.

For our considerations, the details of the FFT are actually not crucial. They are only intended to serve as information. The actual execution of such an algorithm is already included in the package functions in Python and can be applied by us very easily and quickly.

In Listing 3.16 you will find an application example of the Fourier transformation on a simple combined oscillation. We have also added the code that represents the transformation. With this example, you can quickly examine transforms of various input data yourself. In many cases, the Fourier transform can provide a new perspective that improves the view of the data. This is the case even when there are no obvious, clean oscillations in the data. Often, it is then other recurring patterns that can be well recognized in the frequency space.

Listing 3.15 Fast Fourier Transform (FFT)

```python
import matplotlib.pyplot as plt
import numpy as np

# Definition der Frequenzen in Hz
nu1 = 1
nu2 = 4
nu3 = 10

A1 = 1
A2 = 0.3
A3 = 0.15

# Beispieldaten mit diesen Frequenzen
final_t = 20*np.pi
dt = final_t/2**14
t = np.arange(0,final_t,dt)  # in s
x = A1*np.exp(1j*2*np.pi*nu1*t) \
    + A2*np.exp(1j*2*np.pi*nu2*t)\
    + A3*np.exp(1j*2*np.pi*nu3*t)\

# Berechnung der FFT
fft = np.fft.fft(x)
freq = np.fft.fftfreq(x.size, d=dt)

# Normalisierung
fft = np.abs(fft)
fft /= np.max(fft)

# Darstellung
fig, (ax1,ax2) = plt.subplots(2,1,figsize=(16,9))

ax1.plot(t, np.real(x),'k', linewidth=2.0)
ax1.set_xlabel('t / s', fontsize=20)
ax1.set_ylabel('x(t)', fontsize=20)
ax1.set_xlim([0,np.pi])
ax1.tick_params('both',labelsize=18)
ax1.tick_params('both',labelsize=18)

ax2.plot(freq[0:8191], fft[0:8191],'k', linewidth=2)

ax2.set_xlim([0,12])
ax2.set_xlabel('Frequenz freq / Hz', fontsize=20)
ax2.set_ylabel('FFT[x(t)]', fontsize=20)
ax2.tick_params('both',labelsize=18)
ax2.tick_params('both',labelsize=18)
```

Listing 3.17 finally shows the application of the Fourier transform to our engine example. For this, we use the intermediate state reached in Listing 3.17 (the variable `newData`) and calculate an FFT for each data vector.

Listing 3.15 Applying the FFT to the engine example

```
for i in range(0,500):
    fft = np.fft.fft( (newData[i]-np.mean(newData[i]))/np.sum(
        newData[i]))
    fft = np.abs(fft)
    fft /= np.sum(fft)
    if fft[8]>0.015:
        myColor = 'r'
    else:
        myColor = 'k'
    plt.plot(fft, color=myColor, alpha=0.2)
```

Since the FFT actually calculates complex numbers, all results of `np.fft.fft(x)` are complex: they contain a real part and an imaginary part. For this reason, we deliberately chose the complex notation above. You could extract the individual components of a complex number x via `np.real(x)` or `np.imag(x)` or determine their (real) magnitude via `np.abs(x)`. We used the latter in the example.

The results of this short code are illustrated in Fig. 3.9 and show that a specific frequency stands out. We have deliberately marked this frequency, let's call it v^*, in red during preprocessing. All transformations where we see more pronounced values for v^* correspond to the original data with the easily recognizable oscillation. We thus have an easily verifiable criterion whether an oscillation is present or not.

This example of exploratory data viewing, testing a transformation, and ultimately finding suitable criteria to find a disturbance are often recurring steps when working with process and machine data.

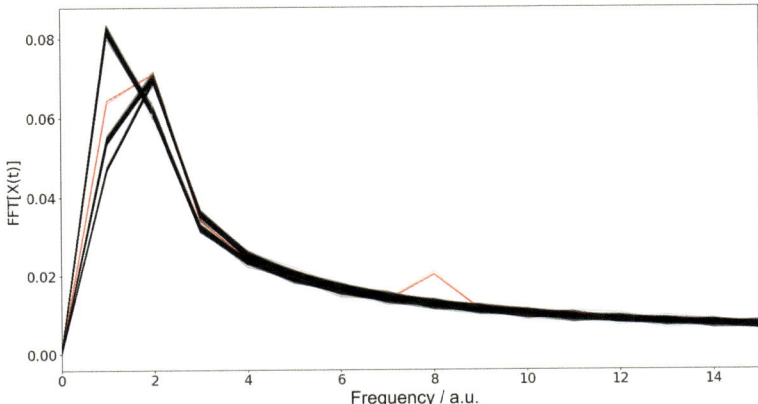

Fig. 3.9 Fourier transform of our motor current example. One curve shows an increase at the dimensionless frequency 8

3.6.4 Wavelet Transformation

In some situations, an oscillatory behavior only occurs in a time window. So it is temporally local. However, the FFT represents our data through oscillating functions, and these are temporally continuous. A better alternative is to choose a projection that can reflect this temporal limitation. This is exactly where the so-called wavelets come into play.

Wavelets are temporally limited wave patterns with a width a and a temporal position b. A simple and easily understandable example is the Ricker wavelet. It is the second derivative of the Gaussian function,

$$\psi\left(\frac{t-b}{a}\right) = \frac{d^2}{dt^2} \frac{1}{\sqrt{\pi a^2}} \exp\left(-\frac{(t-b)^2}{a^2}\right). \tag{3.16}$$

The continuous wavelet transformation $\mathrm{CWT}[x(t)]$ uses this wavelet basic form ψ to obtain a width- and position-dependent projection of the function $x(t)$.

> The **continuous wavelet transformation** of a function $x(t)$ with respect to a wavelet $\psi(\frac{t-b}{a})$ is given by
>
> $$\mathrm{CWT}[x(t)](a,b) = \frac{1}{|a|^{\frac{1}{2}}} \int_{-\infty}^{\infty} x(t)\psi\left(\frac{t-b}{a}\right) dt. \tag{3.17}$$

Note the form of the Continuous Wavelet Transformation result, which depends on b and a. Unlike the FFT, which contains the amplitude as a function of frequency in the result, the wavelet transformation provides an amplitude as a function of b and a, thus a 2-dimensional result. In this form, the CWT thus increases the dimension of the input data. Listing 3.18 first demonstrates the creation of artificial sample data to understand the effect of the CWT.

Listing 3.18 Synthetic example data for the wavelet analysis

```
import numpy as np
import scipy.signal as sp

final_t = 5*np.pi
dt = final_t/2**14
t = np.array(np.arange(0,final_t,dt))
x = np.cos(2*np.pi*1.*t) + 3*np.exp(- 1*(t-1))
```

An exemplary application can be found in Listing 3.19, where we also present the 2D representation of the wavelet transformation.

Listing 3.19 Continuous wavelet transformation

```
%matplotlib widget
widths = 1*np.arange(10,200,2)
cwtmatr = sp.cwt(x, sp.ricker, widths)

plt.imshow(np.real(cwtmatr), cmap='nipy_spectral', vmin=0,
           vmax=np.max(np.real(cwtmatr)),
           aspect='auto')
```

We now want to look at the example with the motor currents from the wavelet perspective. For this, we take the previous code example and supplement it in Listing 3.20, so that we can distinguish good and bad cases based on the Fourier transform. This serves for illustration in this case and is a good example of how the methods can be used in combination.

Listing 3.20 Splitting engine data for the analysis with a CWT

```
Bad = []
Good = []
for i in range(0,500):
    fft = np.fft.fft( (newData[i]-np.mean(newData[i]))/np.sum(
        newData[i]))
    fft = np.abs(fft)
    fft /= np.sum(fft)
    if fft[8]>0.015:
        myColor = 'r'
        Bad.append(newData[i])
    else:
        myColor = 'k'
        Good.append(newData[i])
    plt.plot(fft, color=myColor, alpha=0.2)
```

We now have clearly defined sets for good and bad cases. This is a state that we must always bring about for classification tasks. If we have access to such a divided data set, the subsequent application of a classification algorithm is often simple and without problems. The prerequisite, of course, is that a criterion for distinction can really be found.

We use both sets, Good and Bad, in Listing 3.21, to determine the transforms for individual cases of these sets. The results of this code are summarized in Fig. 3.10. A criterion for the detection of the bad case can also be found in the CWT result.

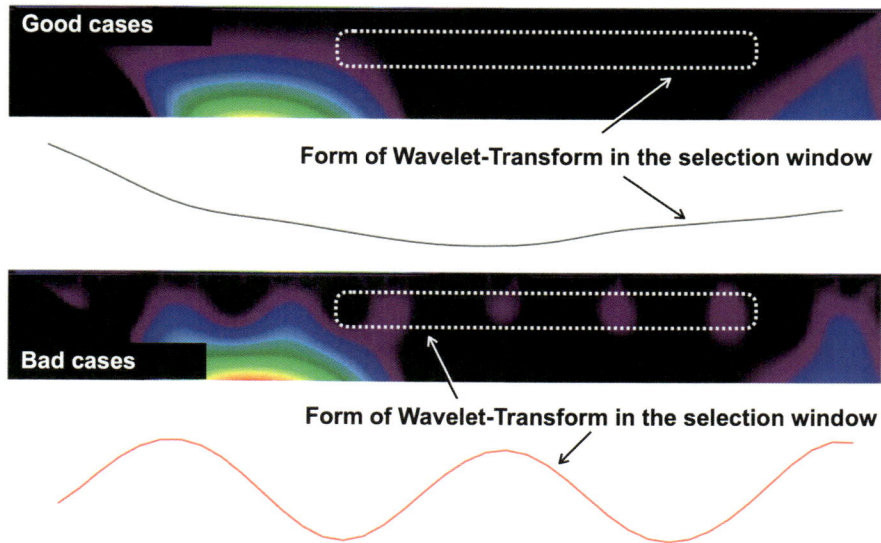

Fig. 3.10 Wavelet transformation of a case from the good set (top) and a case from the bad set (bottom). By cutting out a suitable area in the transforms, a criterion for the set can be derived

Listing 3.21 Continuous Wavelet Analysis using the example of engine data

```
import scipy.signal as sp
widths = 0.1*np.arange(1,100,.1)

goodCwtmatr = sp.cwt(Good[2], sp.ricker, widths)
badCwtmatr = sp.cwt(Bad[2], sp.ricker, widths)

fig, (ax1,ax2,ax3,ax4) = plt.subplots(4,1)
ax1.imshow(np.real(goodCwtmatr), cmap='nipy_spectral', vmin
    =0.0,
            vmax=0.01,
            aspect='auto')
ax2.plot(np.real(goodCwtmatr[300,50:90]),'k')
ax3.imshow(np.real(badCwtmatr), cmap='nipy_spectral', vmin
    =0.0,
            vmax=0.01,
            aspect='auto')
ax4.plot(np.real(badCwtmatr[300,50:90]),'r')
ax1.axis(False)
ax2.axis(False)
ax3.axis(False)
ax4.axis(False)
```

However, detection is more difficult than with Fourier analysis. It even seems to be a disadvantage to use the CWT, as the dimensionality of the evaluation increases. In fact, the benefit of transformation for the recognizability of a separating property is at the heart of almost every exploratory work.

3.7 Quantification of Stochastic Properties

3.7.1 Histograms

We know some selected distributions from the previous chapter, but how can we extract these distributions from our data? Our goal is not only to have a good understanding of the process. Rather, we want to determine further characteristics from the data, with which we can reduce the dimension of the data or which can help a machine learning method to derive models from the data.

In this context, knowledge of the distribution is an important tool. Through the analytically calculable parameters of the distributions, we can define limit values, control the correctness of processes, and distinguish the quality of products. Therefore, we are interested in being able to extract the probability distribution from the data, and this is most easily done through histograms.

We determine through a histogram how often a specific value of a variable occurs. Sensibly, we choose for the frequencies contiguous ranges I_k, intervals that cover a certain range of values of the variable. To create a histogram, we count how often our variable passes through the different ranges I_k.

We say x_i lies in the interval $I_k =]x_{k,min}, x_{k,max}]$, $x_i \in I_k$, when $x_i < x_{k,max} \wedge x_i > x_{k,min}$ is fulfilled. We interpret I_k as a set. We refer to the discrete representation of x_k against the power of all sets $|I_k(x_k)|$ as a **histogram**.

In Listing 3.22 we show how to create a histogram yourself. This is intended for illustration. Our tools in Python offer simpler and faster ways to get this information.

Listing 3.22 Continuous Wavelet Analysis using the example of engine data

```
import matplotlib.pyplot as plt
import numpy as np

x = 2*np.random.randn(1000)
k = range(-10,10,1)
Ik = np.zeros(len(k))

for i in range(0,len(x)):
    for j in range(0, len(k)):
        if x[i] <= k[j]+1 and x[i]>k[j]:
            Ik[j]+=1
```

By using suitable library functions, here `plt.hist()`, the program code for a histogram is reduced to one line in the code 3.23.

Listing 3.23 Calculating a histogram using Matplotlib

```
plt.hist(x, bins=20)
```

It leads to the same goal and is less laborious. With the parameter `bins` we determine in both cases how many intervals we want to use and ultimately, how finely the histogram resolves the different values of the variable.

Histograms are extremely effective in reducing the dimension of the data. In the above example, a histogram maps 1000 data values to an interval of 21 numbers. The histogram primarily captures the stochastic properties of the variable.

3.7.2   Identification of the Probability Distribution Using Kernel Density Estimators

Now that we know how to create histograms, we should return to a recurring theme, namely probability. It is important to understand that histograms represent how likely a certain value of a variable is. From our histogram example, we infer that the $x = 0$ occurs particularly frequently, and is therefore more likely than other values. Numbers greater than 100 are completely unlikely.

> The **histogram** $\mathcal{H}(x; b)$ is a quantitative, discontinuous, and non-normalized representation of the **probability density** $p(x; \mu, \sigma, \nu, \kappa)$ of a stochastic process x.

In inductive statistics, one systematically investigates the gain of information from samples. From here come methods such as the χ^2 test or the Anderson-Darling test, with which we can calculate how well our data is described by a certain distribution.

Although the histogram can be very helpful, it has a crucial disadvantage: it is discontinuous. The division into intervals segments the histogram into fixed blocks. However, the technical process that produced the data and is thus responsible for the histogram does indeed have a continuous probability density.

At this point, the Kernel Density Estimator helps us. The term kernel is used in the field of statistics in various contexts. We will consider it here as the distribution function centered around zero.

> A **Kernel Density Estimator** is a continuous estimate of an unknown distribution. It approximates the unknown distribution by a sum of known distributions.

Let's consider an example with synthetic data in Listing 3.24:

Listing 3.24 Synthetic data for a kernel density estimation

```
import matplotlib.pyplot as plt
import numpy as np

d1 = np.random.normal(loc=70, scale=15, size=300)
d2 = np.random.normal(loc=20, scale=5, size=700)
data = np.hstack([d1,d2])
data = data.reshape((len(data), 1))

plt.hist(data,width=3, bins=50)
plt.show()
```

In this listing, we have drawn events from two Gaussian distributions and histogrammed them. The resulting histogram is shown in blue in Fig. 3.11. The kernel

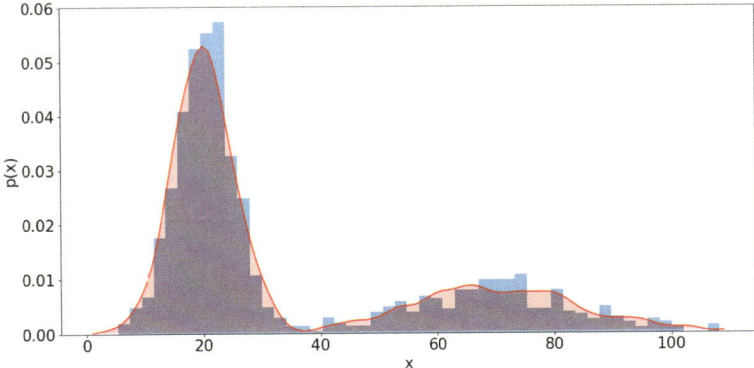

Fig. 3.11 Scaled histogram (blue) of the input data and overlay of the continuous approximation for $p(x)$ (red) using a kernel density estimator

density estimator uses this input data and approximates a continuous combination of distribution functions. In the following code in Listing 3.25, we assume a Gaussian kernel.

Listing 3.25 Kernel density estimation using Scikit-Learn
```
from sklearn.neighbors import KernelDensity
model = KernelDensity(bandwidth=3, kernel='gaussian')
model.fit(data)
```

The visualization of the result is fully executed in Listing 3.26 and shows the code that generated the red distribution and the histogram in Fig. 3.11.

Listing 3.26 Visualization of the kernel density estimation results
```
x = np.asarray([i for i in range(1, 110)])
x = x.reshape((len(x), 1))
y = np.exp(model.score_samples(x))
plt.hist(data, bins=50, density=True, alpha=0.5)
plt.fill(x[:], y, 'r', alpha=0.2)
plt.plot(x[:], y, 'r', alpha=0.9)
plt.tick_params('both', labelsize=20)
plt.xlabel('x', fontsize=20)
plt.ylabel('p(x)', fontsize=20)
plt.show()
```

In line 3 of this listing, an exponential function is used and applied to the random draw from the kernel density model. This is necessary because the function `fit` performs the approximation with logarithmic variants of the distribution.

The quantification of uncertainties is a distinct field of study and is important for the areas of multiparticle physics and the dynamics of complex systems. Here, more extensive approaches are used. For further reading, we recommend the works of P. J. van Leeuwen et al., who use particle filters [4, 6] in the context of meteorological and oceanographic models, as well as the ensemble Kalman assimilation approaches, as discussed by M. Jardak and O. Talagrand in [3].

3.8 Determining Key Figures from Data

3.8.1 Statistical Key Figures

One of the most commonly used concepts for processing large amounts of data is the reduction of dimensions to statistical key figures. Here, the aim is to capture a high-dimensional data set with a few characteristic values. Expected value and moments are suitable measures in many cases to effectively reduce data. But they also serve as target values for learned models.

In some process databases, they even go so far as to only store very condensed key figures of time series, e.g., mean, min, and max. This saves the storage of large data sets from the outset. The sensor delivers all values, but only these few key data are retained. Unfortunately, this step also loses information. Especially the dynamics so important for time series, which is mainly expressed in the gradients and oscillations during the data course, cannot be captured by the distribution parameters.

Example: Acceleration, Deceleration, and Oscillation
Imagine a motor whose speed you first increase from 0 revolutions per minute upwards, then let it fluctuate between two values, and finally decelerate back to 0 revolutions per minute. For all three scenarios, you can get the same maximum, minimum, and average values and in principle no longer distinguish between the actual states. If you find that most faulty states occur during deceleration, then it is not enough to only capture the statistical key figures. ◄

A machine start-up situation would therefore possibly no longer be distinguishable from the braking situation in terms of data. The example makes it clear that there are processes in which the complete evaluation of a time series is necessary and not just the sole consideration of statistical key figures. This problem is investigated for various scenarios in the steel industry in the works of Arnu et al. [1] and Souza et al. [2].

3.8.2   Process-Oriented Key Figures

In principle, the raw data of a dynamic process should also be fully recorded. So first use the Nyquist criterion to determine the most sensible sampling rate for your data. Since we have already seen that statistical quantities alone are not necessarily suitable for always describing the process well, we would therefore like to give you a recipe to sensibly expand your data collection. For this, we focus on characteristic key points of the actual process and determine them in the following way:

- Determine the statistical characteristics of the time series, but also include the temporal positions at which the minimum, maximum, and mean are reached.
- Try to determine the most important frequencies via the FFT and store them. If this is not possible, reduce the frequency space to 4 to 5 frequency ranges and store the averages in these frequency bands.
- Define characteristic points in your data. In time series, always (t_i, x_i). These points are individual and problem-related, but they can be an enormously important tool for formulating target variables. These points can also be

maxima, minima, or turning points in the time series. It is important to jointly capture t and x_i.

- Determine relationships between the characteristic points that are meaningful for your process:
  - Is there a characteristic difference between the maximum and minimum?
  - Do the maximum and minimum have a special relationship to each other?
  - Is the distance of the temporal position of the maximum from the start point of the time series relevant?
- Where are the extreme points of the slope reached? Use the zeros of the derivative to find these points and also store the values t_i and $\frac{dx_i}{dt}(t_i)$.

We want to briefly demonstrate this using our example of the motor current:

Example: Process-oriented Key Figures of the Motor Current Example
Consider the course of the data in Fig. 3.2. We use the variable $x(t)$ to describe the course of a curve. The following points are extracted here:

1. The position $x(t = 0)$ is a process-relevant offset.
2. The slope at the point $x(t = 10)$ must be as identical as possible in all curves.
3. The position t_{Max} must be synchronous for all curves. The maximum is therefore relevant, both in its height and in its temporal position.
4. The position and height of the minimum of each curve are also characteristic and different for each curve.
5. The FFT frequency bands are characteristic. With 5 intervals in the frequency space, the disturbance frequency of the anomaly is also captured.
6. The end position of the curves can also be included as a point. ◀

The consideration of process-oriented key figures is therefore helpful. In the present example, these would be about 10 values and they would be completely sufficient to capture the entire problem.

Summary
Before we can deal with specific learning methods, we had to create the basics in this chapter to prepare data specifically for so-called training. The work involved in preparing the data, especially in cleaning and considering various normalizations, should not be underestimated in practice.

We have seen the step of triggering, which is necessary for repetitive problems. Then we considered various transformations that can change our view of the data and help learning methods to train faster and more robustly. With the transformations, aspects of the data became visible that were sometimes not or only difficult to see in the original data.

Building on the mathematical foundations of the previous chapter, we discussed histograms and kernel density estimators as tools to extract probability distributions from data. They close the circle between theoretical ideas and practical evaluation.

The preparation steps have one goal, namely to sharpen the view of the essentials in the data. This is achieved by omitting the irrelevant or by focusing on those properties that contain the relevant information. Fourier transformation, wavelet transformation or mapping to distributions are capable of reducing complex data to a few characteristic numerical values.

We will later get to know the unsupervised learning methods in Chap. 5. Here, the principal component analysis (PCA) and the autoencoder are well-known tools for effectively reducing the dimension of data. They represent, like the methods of this chapter, procedures that are also suitable for preparation.

Chap. 6 and 7 will finally pick up the preprocessing again under the aspect of a possibly automatic process chain. If a digital system independently makes the decision which preprocessing is best suited, we finally find ourselves in another area of artificial intelligence: knowledge available for the machine.

Tasks

3.1 How can you generate NaNs in a dataset yourself?

3.2 Program the interpolation from equation (3.1)!

3.3 Determine the convolution of the functions $f \circ g$ with

$$f(x) = \operatorname{sech}(x) \tag{3.18}$$

and

$$g(x) = 2 * \delta(x - 10) + 7 * \delta(x - 80). \tag{3.19}$$

Present the result graphically! Here, the δ function is used, which is defined over

$$\delta(x - a) = \begin{cases} 1 \text{ für } x = a \\ 0 \text{ sonst,} \end{cases} \tag{3.20}$$

3.4 Approach the result of task 3.2 with the kernel density estimator. Use kernels that we have not explicitly addressed in the text: a) Cauchy kernel, b) Epanechnikov kernel, and c) Picard kernel. What differences do you notice?

3.5 What are the process-oriented, characteristic features of the triggered example in Fig. 3.6?

References

1. D. Arnu, E. Yaqub, C. Mocci, V. Colla, M. J. Neuer, G. Fricout, X. Renard, C. Mozzati, and P. Gallinari, *A reference architecture for quality improvement in steel production*. Springer, 2017.
2. A. d. M. Souza, D. Arnu, F. Temme, E. Klapic, R. Klinkenberg, M. J. Neuer, X. Renard, P. Gallinari, C. Mozzati, C. Mocci, and G. Fricout, „Data mining and modelling," *Steel Times International*, 2018.
3. M. Jardak and O. Talagrand, „Ensemble variational assimilation as a probabilistic estimator – part 1: The linear and weak non-linear case," *Nonlinear Processes in Geophysics*, vol. 25, no. 3, pp. 565–587, 2018. [Online]. Available: https://npg.copernicus.org/articles/25/565/2018/
4. P. J. V. Leeuwen, „Aspects of particle filtering in high dimensional spaces," *Lecture Notes in Computer Science*, vol. 8964, pp. 251–262, 2015.
5. M. J. Neuer, A. Quick, T. George, and N. Link, „Anomaly and causality analysis in process data streams using machine learning with specialized eigenspace topologies," in *Proceedings of ESTAD 2019*, 2019.
6. S. Pathiraja and P. J. van Leeuwen, „Multiplicative non-gaussian model error estimation in data assimilation," 2021.

Chapter 4
Supervised Learning

Keywords Supervised learning · Adaptive filters · Evolutionary algorithms · Neural networks · Decision trees · Gradient descent methods

This chapter discusses concepts of supervised learning. It introduces a general approach to learning methods, which serves as a basis for the following approaches. Starting with a naive, evolutionary algorithm, we discuss the Least-Mean-Squares (LMS) adaptive filter, neural networks, recurrent neural networks, and decision trees in the course of the chapter. Example implementations are provided for each method, giving the reader a starting point for their own projects.

Fig. 4.1 provides an overview of how the following chapters benefit from the content presented here. A central concept is the cost function J. It allows us to formulate learning as an optimization problem. This basic approach will accompany us in Chap. 4 and 5.

The learning methods themselves form the basis for Chap. 6. Chapter 7 shows methods that are applied to a trained model to better understand it.

4.1 Learning

4.1.1 What does Learning Actually Mean?

- **Associative Learning.** In associative learning, the aim is to link two or more events together in the learner. Mathematically, if-then links are to be developed, where

$$X \rightarrow Y \rightarrow Z \tag{4.1}$$

© The Author(s), under exclusive license to Springer-Verlag GmbH, DE, part of Springer Nature 2025
M. J. Neuer, *Machine Learning for Engineers*,
https://doi.org/10.1007/978-3-662-69995-9_4

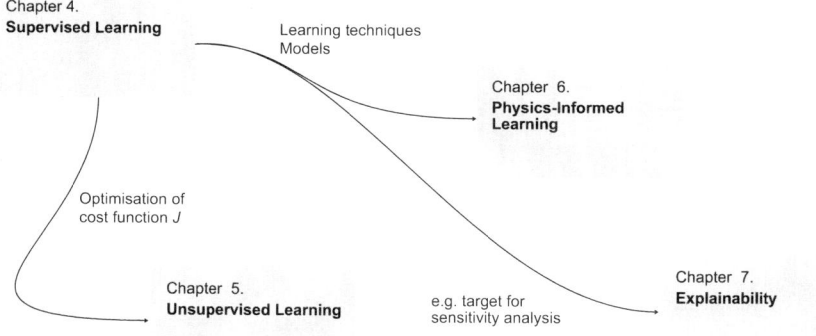

Fig. 4.1 Overview of the relationship of Chap. 4 with the following chapters

means: If event X occurs, then Y also occurs and if Y occurs, then Z follows. **Classical conditioning** is a form of associative learning where a neutral stimulus, e.g. a tone, is linked to an expected reward. Whenever your dog or cat does something you want, you press a clicker and then give a reward in the form of food. This reward is called reinforcement. The animal eventually associates the sound of the clicker with the reward. Hence, this form of learning is also called associative learning.

- **Non-associative Learning.** In contrast to associative learning, non-associative learning deals with exactly one event X—a phenomenon, a noise, or an impression. If this event X occurs often enough, we derive an interpretation from it. The constant noise of water that we gradually get used to, or the decrease of a smell perception that we are exposed to for a long time, are examples of the so-called non-associative **habituation.**

- **Playing.** One of the most impressive and effective learning methods is to play. It is an almost unlimited space for trials and errors, allowing for free development of concept understanding. The ability to play is a fundamental characteristic of intelligent beings.

- **Cultural Learning.** Culture consists of many aspects, with rites, societal rules, laws, and even moral concepts being passed on through culture. This does not always lead to advantageous developments, as the often misguided interpretation of the words "normality" and "naturalness" shows.

- **Imitation Learning.** Role models are people who make a positive impression on us or people we want to emulate. Such emulation is an example of imitation learning. In the animal kingdom, it is observed that older animals pass on certain tricks to young animals to get food. They perform an action and the learners follow it. In the technical field, this is particularly relevant for training robots for production processes, which learn their paths and movement patterns from the user using teach-in techniques.

4.1.2 Various Forms of Machine Learning

Machine Learning can be divided into three forms: supervised, unsupervised, and reinforcement learning.

Supervised Learning deals with the establishment of models f that map input data x to the results d associated with x, $d = f^{(\infty)}(x)$. We assume the availability of d. This gives this type of learning its name supervised: we control towards which result the learning process is trained.

Unsupervised Learning only considers a set of input data. Categorizations or model relationships in the data are unknown. The learning process is now supposed to discover structures in the data. It analyzes whether the data are connected as clusters or to what extent data areas are related to each other.

Reinforcement Learning refers to autonomous agents. These agents have a quantitative goal, their satisfaction. By specifically increasing or decreasing satisfaction, the agent learns whether a decision was good or bad. This type of learning can be trained on online data streams and can be continued continuously. Therefore, it has great relevance for control technology. In fact, we will not return to reinforcement learning in the course of this book, as it does not fit into the framework of this introduction.

4.2 Strategy of Supervised Learning

4.2.1 Algorithmic Consideration

The goal of supervised learning is to map input data x to output data y, so that they represent a known target size d. Often, d is also called a label—in reference to classification problems. We learn by iteratively changing the model f with $y = f^{(i)}(x)$, $f^{(i)} \rightarrow f^{(\infty)}$, so that after the change $f^{(\infty)}(x) \approx d$ results. Here, $f^{(\infty)}$ stands for the converged model, which is realized in practice for very large i or low cost values. To do this, we must be able to evaluate the change of the model f in each iteration.

This evaluation is done via a quality function tefunktion J, often also called quality functional, cost function or simply cost. If J is large, the model is bad, if it is small, the model is good. If you know the quality function for f, you can decide by comparing this quality whether the model has improved, remained the same or even worsened when changed. The strategy can be divided into the following steps:

- **Setting up an initial model f.** This model is based on initial conditions for the learning parameters and is accordingly bad. It will generate an arbitrary mapping that has no relation to the desired reference results d at first.
- **Calculation of $J(f)$.** Here, the quality of the current model is determined. It then serves as a reference value for the next steps.

- **Changing the model.** We now change f by $f \rightarrow \tilde{f}$. How f changes is dictated by J, because we want J to decrease. This step is critical for learning. The parameters used in the model f are specifically adjusted for the new model $\tilde{f}$.
- **Iteration of the previous step.** We iterate the last step until the costs J have dropped to an acceptable value. This is equivalent to a correspondingly good mapping quality by f.

> Supervised learning of a model f describes the iterative change from $f^{(i)}$ to $f^{(i+1)}$, so that
>
> $$f^{(i)} \mapsto f^{(i+1)} \Rightarrow J(f^{(i+1)}) \overset{!}{<} J(f^{(i)}) \Leftrightarrow \min_i J(f^{(i)}). \qquad (4.2)$$

This simple, universal scheme can be used as the basis for many algorithms. They differ in the type of change step or the choice of cost function J—but almost always follow the procedure outlined in (4.2). Learning algorithms are therefore always optimization problems. This is also evident in (4.2): We are looking for the minimum of J given models $f^{(i)}$. We want to introduce some selected methods in the course of this chapter and briefly summarize them here:

- **Evolutionary algorithms.** In evolutionary algorithms, J corresponds to the reciprocal of a fitness. The variation of the model f occurs through random throws. Whenever J decreases from one step to the next, i.e. the fitness increases, the change is retained. If J increases, the result worsens and the new parameter choice is discarded. Afterwards, one continues to iterate, based on the last good variant. Closely related to this are the genetic algorithms.
- **Least-Mean-Square algorithm.** In the LMS adaptive filter, filter coefficients dynamically adjust so that a target value is achieved by the filter. J is chosen here as the mean square deviation between the target value and the current model result.
- **Neural networks.** Neural networks are nonlinear mappings for which various cost functions J are possible. They train according to the same concept, but mathematically already know the best gradient of J with which they need to change the internal parameters. This method is called backpropagation, and will occupy us in the next sections.
- **Decision trees.** Decision trees search for optimal features in the data to represent a dataset as well as possible. The tree identifies the variables that have the strongest influence on a decision. The tree splits the decision into branches based on these variables and their values. Here too, the quality of the mapping is assessed with the help of a cost function J. It determines the deviation from the target value and specifically adjusts thresholds for the splitting of branches.

4.2.2 Classification and Regression

Goal of supervised learning is the creation of a model. This model can represent our input data x at different levels of granularity.

- **Classification.** If the input data should be assigned to a class, then we only distinguish between a few results in the result space. Classifications have discrete output data.

> **Example: Traffic Light and Motorcycle**
> An image recognition should recognize motorcycles and traffic lights in videos. The result set is $\Omega = $ ("Motorcycle", "Traffic Light") and $d \in \Omega$. ◀

> **Example: Alarm Classification**
> A measurement on a capacitor is examined for voltage peaks. The waveform reflects these peaks. Using learning methods, it is decided whether the signal course is dangerous or not. We use the voltage course as input data and for d we need assignments $d \in \{0, 1\}$ where 0 stands for harmless and 1 for dangerous. This is a canonical classification situation. ◀

- **Regression.** A regression model replicates a complete functional course. The output values are continuous over a defined set of values. Regression models require more data and a minimum resolution in the labels or target values on which the method should be trained.

> **Example: Probability of Danger**
> We continue our last example and reformulate it: The evaluation of the signal should not only distinguish between the two categories 1 "Danger" and 0 "No Danger", but rather output a risk value that, for example, lies between 0 and 1. With a result of $y = 0.8$ a 80 % risk is present. This is a classic example of a regression. ◀

4.2.3 Training Data and Test Data

We now define how we want to make the system change according to the existing data and how we calculate the cost function J. At this point, the knowledge of the desired output value d is important. By comparing d and y we can determine the deviation of the current prediction.

To train a supervised learning model, we need sufficient and suitable training data. Sufficient, because we often have to iterate and change in (4.2). Suitable data, as these must always consist of tuples $(x, y \rightarrow d)$, where x represents the input data and d the desired, ideal mapping result. Subsequently, we will use y and d synonymously.

To establish a learning procedure, we need two dedicated data sets, the **training data set,** $x_{i,\text{Train}} \in X_{\text{Train}}$, $y_{i,\text{Train}} \in Y_{\text{Train}}$ and the **test data set,** $x_{j,\text{Test}} \in X_{\text{Test}}$, $y_{j,\text{Test}} \in Y_{\text{Test}}$. The training data set consists of $|X_{\text{Train}}| = |Y_{\text{Train}}| = N$ data series, the test data set contains $|X_{\text{Test}}| = |Y_{\text{Test}}| = M$ elements. In the code, we always refer to these sets as **Xtrain, Ytrain** and **Xtest, Ytest.**

Since we need more data for training a model, the training data set should contain significantly more data than the test data set, $N \gg M$. Often in practice, a division of 90 % training data and 10 % test data is used. But why are these sets so important? On the one hand, we want to know if the trained learning procedure also works on data that it has never evaluated during training. Only in this way can we ensure in principle that these data will be correctly predicted by the procedure for new data.

The training and test set should also statistically contain all possible cases, scenarios, and classes that you want to learn. If, for example, your training set does not cover all classes of the test set, you would receive distorted test results, since an important part could not be trained at all.

Golden rule for training learning methods: Elements of the test data set must never enter the training!

If you violate the above rule, your test results will be almost perfect without ever being able to transfer the trained model, i.e., apply it meaningfully to new data. This is one of the most common mistakes encountered in practice. We will later learn about the concept of cross-validation, which cleverly mixes the two sets and thus excludes this error.

In practice, it is a great difficulty to set up these data sets for the application of learning methods at all. Not all process databases are designed with the aim of supporting learning methods. Many database structures have grown over the years, and compiling these two data sets takes a lot of time. We have to ask ourselves the following questions: Do we actually have the required labels d on the data side? Are there enough data in total to train a learning method? These questions must be clarified in a planning manner before the development of a machine learning program.

Another critical point is the process change. Let's assume our learning method is supposed to generate a model f for a process and we actually had enough data. If the process itself has changed while the data was being recorded, the model is useless. We would first have to divide the data into "before the change" and "after the change" split. As you can see, this reduces the amount of data again. All these difficulties are often encountered in the industry.

4.3 Evolutionary Learning

4.3.1 Idea of Evolutionary Change

In evolutionary algorithms, we first talk about populations, in reference to biology. A population describes the parameters $\boldsymbol{\pi} = (\pi_i)$ of our model f, which maps x to y,

$$y = f(x; \boldsymbol{\pi}). \tag{4.3}$$

Our goal is that for a set of input data (x_j, d_j) the parameters $\boldsymbol{\pi}$ in f are determined in such a way that $f(x_j; \boldsymbol{\pi} = \boldsymbol{\pi}^*) = d_j \ \forall j$. Here, j numbers our training set and $\boldsymbol{\pi}^*$ describes the ideal parameter set. We only discuss the one-dimensional case here, the method can easily be extended to any dimensions.

The fitness F of the Population Π with $\pi_i \in \Pi$ is given by

$$F = \frac{1}{J} = \frac{1}{\sum_j (d_j - y_j)^2}. \tag{4.4}$$

The more d and y match, the higher the fitness F and the lower the cost function J.

What we still lack is the change of $\boldsymbol{\pi}$. And here we choose a not very efficient, but extremely instructive variant: we randomize $\boldsymbol{\pi}$ until the mapping fits.

4.3.2 Implementation of a Simple Evolutionary Algorithm

The following code example shows a simple evolutionary algorithm that tries to find the best result using only random numbers. It is instructive because many brute-force approaches find a solution in a similar way. However, this approach is also inefficient, as one has to wait a long time for a good solution to emerge.

We divide the implementation of an example into four sections. In section (1), we generate artificial training data. They simply represent a normal parabola that we can easily check. Next, we define the function f in section (2) and insert an array p with parameters p[0], p[1] and p[2]. In our example, the function describes a 2nd degree polynomial of the form

$$f(x, \boldsymbol{\pi}) = \pi_0 + \pi_1 x + \pi_2 x^2. \tag{4.5}$$

Listing 4.1 A simple evolutionary algorithm

```
 1  import numpy as np
 2
 3  # (1)  Input data
 4  x = np.arange(0,5,0.1)
 5  d = np.arange(0,5,0.1)**2
 6
 7  # (2)  Function  f
 8  def f(x,p):
 9      return  p[0]+p[1]*x+p[2]*x**2
10
11  # (3)  Definition of fitness
12  def Fittness(d):
13      j = 0
14      for i in range(0,len(d)):
15          j+= (d[i]-f(x[i],p))**2
16      return 1/j
17
18  # (4)  Loop for optimizing J and F
19  maxF = 0
20  F = 0
21  while F < 3:
22      p = np.random.random(3)
23      F = Fittness(d)
24      if F > maxF:
25          maxF = F
26          print(maxF)
```

In section (3), the fitness is calculated. We use the square deviation from the target value d. Finally, section (4) contains the iteration steps. π is rolled again each time. With each roll, a new parameterization is given for which a fitness is calculated. As soon as a fitness value is reached that is greater than $F = 3$, the iteration stops. The specific numerical value was found here through tests.

The result of this learning process can be checked well, because for the chosen example, p[0] and p[1] should be as close to 0 as possible and p[2] should be 1. In practice, they find values around 0.04 for the first two parameters and 0.98 for the last one at the chosen limit. The result is then close enough to the data.

4.3.3 Caching the Best Change

In the next code example, we change the view from the fitness F to a view of the cost function J. In addition, we remember the change deviation that brought us the last success in each step. If it was successful, then another step in this direction cannot be wrong.

Listing 4.2 Variation of the evolutionary algorithm

```
11  # (3)   Definition of fitness
12  def Fitness(d):
13      j = 0
14      for i in range(0,len(d)):
15          j += (d[i]-f(x[i],p))**2
16      return 1/j, j
17
18  # (4)   Loop to optimize J and F
19  J = 1000
20  minJ = 1000
21  p_old = [0,0,0]
22  deviation = [0,0,0]
23  while J > 0.1:
24      p = np.random.random(3) + np.array(deviation)
25      F, J = Fitness(d)
26      if J < minJ:
27          minJ = J
28          deviation = 0.0001*J * (p-p_old)
29      p_old = p
```

But how can we analyze the effect of our intervention in the code? We would need to find an additional quality criterion, one with which we can compare the two variants. To do this, it makes sense to put the number of successful changes in relation to the total number of attempts. In Listing 4.3, we have changed the loop again. We have introduced two counters: counter which increments each time a change step was good, and allCounter which records every iteration step.

Listing 4.3 Variation of the evolutionary algorithm

```
11  allCounterStorage = []
12  counterStorage = []
13
14  for i in range(0,10):
15      J = 1000
16      minJ = 1000
17      p_old = [0,0,0]
18      deviation = [0,0,0]
19      q = 0
20      a = 1
21      b = 0
22      allCounter = 0
23      counter = 0
24
25      while J > 0.1:
26          p = a * np.random.random(3) + b*np.array(deviation)
27          F, J = Fitness(d)
28          allCounter += 1
29          if J < minJ:
30              counter += 1
31              minJ = J
32              a = 0
33              b = 1
34              deviation = 0.00001*J * (p-p_old)
35              q = 0
36          p_old = p
37          q += 1
38          if q > 10:
39              a = 1
40              b = 0
41              q = 0
42      allCounterStorage.append(allCounter)
43      counterStorage.append(counter)
```

Then the `while` loop was embedded in an outer `for` loop, which realizes exactly ten runs. The reason for this is to generate a little more statistics than a single run for our analysis. The variables `allCounterStorage` and `counterStorage` store the respective counter states for each of the ten runs.

Furthermore, a new form of calculation has been implemented. p is first generated randomly. The prefactors are $a = 1$ and $b = 0$, so that really only chance is used. If an improvement of J occurs, we set $a = 0$ and $b = 1$. The deviation is added ten times in a row. After that, $a = 1$ and $b = 0$ are set and a new random value is used.

We can easily compare all variants. It is left to the reader to modify the listing in such a way that only the initial variant from Listing 4.1 is used, which would correspond to a permanent $a = 1$ and $b = 0$. The prefactors in the loop can also be adjusted.

As soon as the ratio of good changes (`counter`) to the total number of iterations (`allCounter`) becomes larger, the algorithm has been formulated more efficiently.

The variation in which we remember the best path of change and follow it for some time is superior to pure chance. This is not surprising. Above all, it shows us that the Least Mean Squares strategy is effective in focusing on the rate of change of J.

4.4   LMS Adaptive Filter

4.4.1   Theory of the LMS Algorithm

The next learning algorithm we want to discuss is the Least-Mean-Squares adaptive filter. Fundamental works on this topic come from Haykin et al. [6] and Widrow et al. [15], where in addition to filter-theoretical derivations, stability investigations and convergence criteria are described. The adaptive filter has also been successfully used for learning language, as shown by Koike in 1997 [9].

As the name already suggests, this learning algorithm comes from the field of filter theory. Let's assume we have a vector $a \in \mathbb{R}^N$ of dimension N. Please imagine, without loss of generality, that the entries of this vector are completely random at the beginning. Our input data is another vector $x \in \mathbb{R}^N$ of the same dimension. Now, a acts like a filter on this input to produce a scalar result $y \in \mathbb{R}$—our output data –,

$$a^T x = y. \tag{4.6}$$

The effect of a on x is formulated here in (4.6) as the scalar product of the two vectors. We use the symbol T for the transposition of the vector a. We want a to "learn" to map x to a defined, desired target number d. We use the symbol d in reference to the English literature, which also likes to refer to this value as *desired value*. So we ask the question: "How does a have to change for this?" This change is implemented in the form of several iteration steps, which we capture using the notation $a^{(0)}, a^{(1)}, \dots a^{(i)}$, where i denotes the i-th iteration. If we use a transposition here, we show this with $a^{(i),T}$ on. The first step is to find out how far d deviates from the current result y, i.e., to calculate the error $\varepsilon^{(i)}$,

$$\varepsilon^{(i)} = d - a^{(i),T} x^{(i)}. \tag{4.7}$$

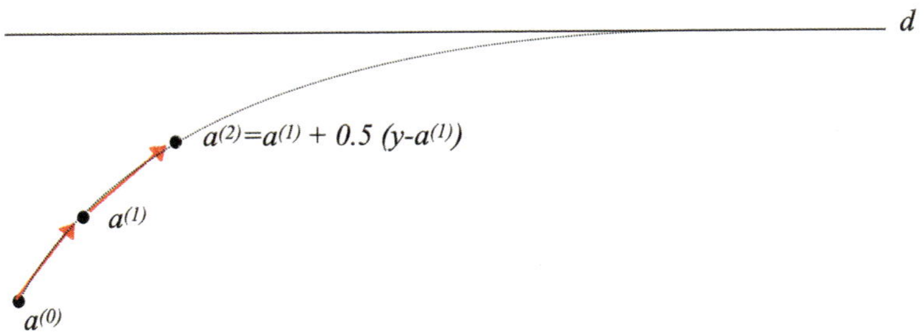

Fig. 4.2 Various iterations of $a^{(i)}$ and approximation to the preset value d. A learning rate of 0.5 was chosen as examples

The error also refers to the current iteration and is marked accordingly, as it changes with every change of a. We can read a lot from this error. If the product $a^{(0),T}x^{(0)}$ is, for example, larger than d, then $a^{(0)}$ is too large and needs to be reduced. The same applies in reverse. The error therefore contains the instruction on how strongly and in which direction we should change $a^{(0)}$.

We need a way to find out this instruction. To do this, we make the following demand: With each iteration, the square of the error should become smaller. Formally, one writes (Fig. 4.2),

$$J^{(i)} = \left(d - a^{(i),T}x^{(i)}\right)^2, \tag{4.8}$$

and demands

$$\lim_{i \to \infty} J^{(i)} \stackrel{!}{=} 0. \tag{4.9}$$

If J is to move quickly towards such a minimum, we only need to determine the negative gradient of J with respect to $a^{(0),T}$. The gradient, which we write over ∇J with the Nabla operator, always points in the direction of the steepest increase, consequently the negative gradient points in the direction of the steepest descent. So we derive (4.8) according to the individual entries of a,

$$-\nabla_{a^{(i),T}}J = -\nabla_{a^{(i),T}}\left(d - a^{(i),T}x^{(i)}\right)^2 = 2\,\varepsilon^{(i)}\,x^{(i)}. \tag{4.10}$$

Please note that we derive over the operation $\nabla_{a^{(i),T}}$ according to the values of a^i and we use the chain rule in this derivation step. We want to change the values of a according to (4.10) to minimize J, and we implement this change in the following equation:

$$a^{(i+1)} = a^{(i)} + \lambda\,\varepsilon^{(i)}\,x^{(i)}. \tag{4.11}$$

It contains a rule for the iteration $i + 1$. The dimensionless factor λ also captures the factor 2 from the calculation in (4.10) and is called **learning rate** . If the learning rate is small, then $a^{(i)}$ only slowly approaches d with each iteration, if it is larger, the approximation is carried out faster. Fig. 4.2 shows the convergence of the method.

Due to its derivation from the square of the error in (4.8), this approach is also called Least-Mean-Squares (LMS) adaptive filter. We will always refer to this as the LMS algorithm from now on.

4.4.2 Implementation of the LMS Algorithm

We now consider how we can use (4.11) in the code. For this, we have created a code example in Listing 4.4. If you run this, y quickly reaches values close to d. We can consider this example as the simplest form of a learning program.

Listing 4.4 Implementation of the adaptive filter in Python

```python
import numpy as np

a = np.random.random(3)
x = [1,2,3]
d = 1.0
eps = d - np.dot(a,x)  # Fehler

Lambda = 0.001 # Lernrate
while eps**2 > 0.0000001:
    a = np.array(a) + Lambda*eps*np.array(x)
    eps = d - np.dot(a, x)

y = np.dot(a,x)
print('y=', y)
```

4.4.3 NLMS Algorithm by Normalizing the Input Variables

In fact, the implementation in Listing 4.4 is not stable. As soon as you use higher values for λ, abbreviated in the code by l, the calculation of y will diverge and not yield a result. It is therefore difficult to define a suitable learning rate for the respective problem so that the algorithm always converges. The reason for this is the influence of x, whose magnitude affects the calculation quadratically over line 10—via the product $\varepsilon\,x$, where ε also contains x once again. From a stability point of view, it makes sense to normalize x by the substitution

$$x \to \tilde{x} = x/(x^T x) \tag{4.12}$$

. From (4.11) the learning rule

$$a^{(i+1)} = a^{(i)} + \lambda\, \frac{\varepsilon^{(i)}\, x^{(i)}}{(x^{(i),T}\, x^{(i)})}. \tag{4.13}$$

is then derived. In the code example, this only leads to a change in line 10, as only the scalar product of x is divided by itself. This line is listed in Listing 4.5.

Listing 4.5 Alternative line 10 for Listing 4.4

```python
a = np.array(a) + l*eps*np.array(x)/np.dot(x,x)
```

While the code in Listing 4.4 represents an implementation of the LMS algorithm, 4.5 is referred to as the NLMS algorithm, where N stands for normalized.

4.4.4 Training

The `while` loop implemented in lines 9–11 continuously applies Eq. (4.13). Each new loop is another iteration. In the code example, we do not explicitly store the individual iterations, but overwrite the value for *a*. The variable *x* remains constant for simplicity and does not change with the iteration.

The process that the `while` loop represents is a concrete example of training a learning procedure. The LMS algorithm and the NLMS algorithm are particularly well suited to understanding the approach of supervised learning programs. The algorithm learns supervised because it has a specification with *d* on which it should train.

4.4.5   Applications and Properties of the Adaptive Filter

The adaptive filter has many technical applications. It allows to dynamically respond to changes and quickly adapt to these changes. Often the target specifications are formulated more complex than in our derivation. However, these are always comprehensible extensions that build on the basic concept above.

> **Example: Speed Controller**
> If you give a target value to a vehicle's cruise control, the speed controller uses the strategy in (4.11) to iteratively adapt to the specification. If the target value was, for example, $d = 50$ km/h and the current speed $v = 20$ km/h, then the acceleration would be large at the beginning and decrease as it approaches the target value through the filter. ◀

However, there are also weaknesses of the algorithm that we would like to briefly explain here.

- Variability of *x*. If *x* changes quickly from iteration to iteration, the basic idea is void and the circumstances are so unfavorable that (4.11) no longer works. If the input values fluctuate more than an adjustment via λ is possible, then the strategy must fail. The filter can no longer keep up with the change in inputs.
- Variability of *d*. This algorithm also has problems if the target size should change.

4.4.6   Hyperparameters of the LMS Adaptive Filter

Obviously, the value of λ plays a crucial role in (4.11). It is the only obvious parameter of the algorithm. A less obvious parameter is the temporal distance of the iterations, i.e., the time that passes between *i* and $i + 1$. It was not addressed in the derivation of (4.11), but it has an impact on the algorithm.

Whenever we define quantities that have an influence on the effect and quality of an algorithm, we speak of **hyperparameters.** The choice of these parameters is crucial for the success of our procedure.

> Factors that we can adjust or choose in a learning procedure and that have a significant influence on the quality of the procedure in terms of learning speed, model quality, and stability are called **hyperparameters.**

4.5 Neural Networks

4.5.1 The Neuron

Neural networks are nonlinear mappings of an input x to an output y. They allow us to learn relationships from data. To understand such a network, we first formulate what we understand by a neuron:

> Let x be a vector of input values, then the function
>
> $$\mathcal{N}_k(x) = f\left[\sum_{i=0}^{N-1} w_{ik}x_i + b\right], \qquad (4.14)$$
>
> is called the k-th **neuron** with the weights w_i, the threshold b and the activation function f.

A neuron thus calculates a state value from N inputs x_i. This is the result $\mathcal{N}(x)$ of (4.14) and is passed on to subsequent neurons. To calculate the state, the inputs are multiplied by weight factors w_i. They reflect the influence of each individual i-th input. We also call the sum formation the input through the network itself and abbreviate this with

$$\text{net}_k = \sum_{i=0}^{N-1} w_{ik}x_i \qquad (4.15)$$

The structure of a neuron is shown in Fig. 4.3.

The bias b is a parameter that can be used to change the effect of the activation function. It shifts it by a constant value. If b is high, it is more difficult to activate the neuron. If it is low, it is easier to activate. Ultimately, the so-called activation function f is applied to the weighted inputs and the threshold. There are many different functions available, which we will discuss in more detail later.

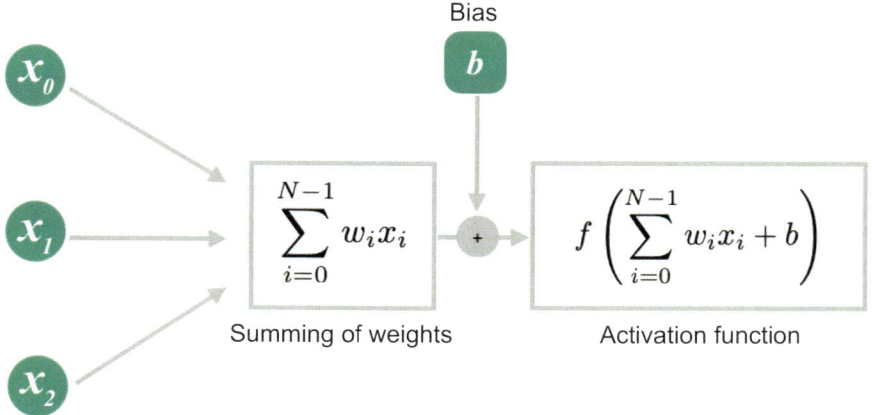

Fig. 4.3   Abstract illustration of a neuron, the smallest component of neural networks

4.5.2   Networking of Neurons

Neurons can be connected together. The activation state of a neuron is used as an input for the next connected neuron. Fig. 4.4 shows a simple network of five neurons. The network shown consists of two **layers,** the input layer with $N = 3$ neurons and the output layer with $M = 2$ neurons. The input layer represents an exception to the above definition, here the excitation of the neurons is not calculated, rather the respective values are the actual input of the network.

In general, the neurons can be linked arbitrarily, but the example shown only connects the neurons of different layers. No connection is used within a layer. Such a network architecture is called a **Multi-Layer Perceptron.** It represents a simple, frequently used form of neural networks.

For this network, we can calculate the output neurons y_0 and y_1. The input neurons are linked to the output neurons via the weights w_{ij}. This layer connection can thus be understood as a matrix, where the row number stands for the target neuron (here 0 or 1) and the column indicates from which original neuron the input comes. In Fig. 4.4, these weights are accordingly drawn onto the connection lines.

From the definition of the neuron, the rule with which we can calculate the excitation of neurons of the k-th layer follows directly,

$$y_k = f\left(\sum_{i=0}^{N-1} w_{ki}x_i + b_k\right),\tag{4.16}$$

which can be written vectorially as

$$y = f(Wx + b),\tag{4.17}$$

with the weight matrix W,

Fig. 4.4 Simple neural
network with three neurons
in the input layer and two
neurons in the output layer

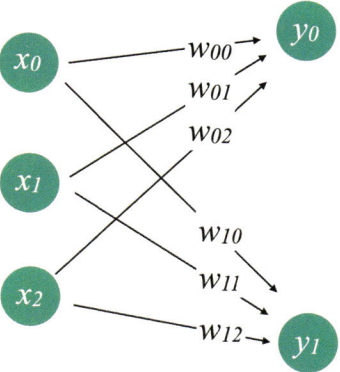

$$W = \begin{pmatrix} w_{00} & w_{01} & w_{02} \\ w_{10} & w_{11} & w_{12} \end{pmatrix}, \tag{4.18}$$

and the threshold vector (engl. *bias*) $\boldsymbol{b}$,

$$\boldsymbol{b} = \begin{pmatrix} b_0 \\ b_1 \end{pmatrix}. \tag{4.19}$$

From this simple case, we extend our discussion to more complex networks. For illustration, a network with four layers is sketched in Fig. 4.5. For each layer transition, there is a weight matrix $\boldsymbol{W}_i$. Layer 0 is the input layer, layer 3 represents the output of the network.

> In a **Multi-Layer Perceptron** with k layers, the input layer $\boldsymbol{x}_0$, the weight matrices $\boldsymbol{W}_k$ and the threshold vectors $\boldsymbol{b}_k$, the values of the subsequent layers can be calculated with $k > 0$ over
>
> $$\boldsymbol{x}_k = f_k(\boldsymbol{W}_{k-1}\boldsymbol{x}_{k-1} + \boldsymbol{b}_k) \tag{4.20}$$
>
> calculate, if f_k is the activation function of the k-th layer. We also refer to this calculation as the forward evaluation of the network.

Along the line in Fig. 4.5, the network is thus evaluated **forward**. One iteratively calculates from $\boldsymbol{x}_0$ the next layer $\boldsymbol{x}_1$ and then continues until the last layer has been determined.

Although we now know how to describe a network abstractly and calculate it from the network input to the output, we have not yet explained how the matrices $\boldsymbol{W}_k$, the threshold vectors $\boldsymbol{b}_k$ and the activation functions f_k can be found.

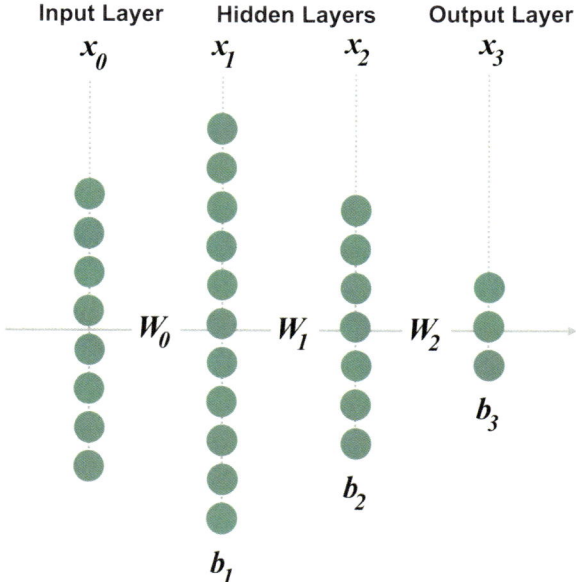

Fig. 4.5 Neural network with multiple layers, weight matrices, and threshold vectors

In the textbooks by Ripley [13] and [1], neural networks are explained with great theoretical depth and for various other forms of realization, so-called topologies. We will limit ourselves here to the exemplary implementation of simple networks in Tensorflow and Keras, as they are most relevant for technical application. For all deeper considerations, the above works are recommended.

4.5.3   Activation Functions

To give you an overview of the possible activation functions, we will now summarize some selected functions. This compilation is for information only, as we can retrieve them later via Python libraries without having to program them ourselves.

- **Sigmoid Function.** The Sigmoid function is one of the most well-known activation functions for neurons. It is given by

$$\text{sig}(x) = \frac{1}{1 + \exp(-x)} \tag{4.21}$$

and has a value range between 0 and 1.

- **Hyperbolic Tangent.** Similar to the sigmoid function in terms of its course, tanh maps the input $x \in \mathbb{R}$ overthe

$$f(x) = \tanh x = \frac{\exp(x) - \exp(-x)}{\exp(x) + \exp(-x)} = 2\operatorname{sig}(2x) - 1 \qquad (4.22)$$

to the value range -1 to 1. Thus, such neurons can take on negative and positive values. We have also listed the relationship with the sigmoid function in the equation.

- **Rectified Linear Unit (ReLU).** An activation function that is successfully used in many current neural networks is the ReLU function:

$$\operatorname{relu}(x) = \begin{cases} x & \text{for } x \geq 0 \\ 0 & \text{else.} \end{cases} \qquad (4.23)$$

It is linear for positive x and 0 for negative x. This fact makes it particularly easy to calculate its derivative, which is always 1 for positive x and 0 in all other cases. ReLU is excellent for hidden layers.

- **Leaky-ReLU.** An important variant of the ReLU function is the Leaky-ReLU. This avoids a training situation where a neuron is trained to death, i.e., it has no gradient to contribute to training for a longer period. The function varies ReLU as follows,

$$\operatorname{leaky}(x) = \begin{cases} x & \text{for } x \geq 0 \\ \varepsilon x \text{ mit } \varepsilon \ll 1 & \text{else} \end{cases} \qquad (4.24)$$

and thus always introduces a positive gradient, even in the range of negative x.

- **Softmax.** Specifically for dealing with the output layer in classification problems, an activation function is needed that allows for probability statements. This function is called Softmax and is defined as follows,

$$\operatorname{softmax}(x_i) = \frac{\exp(x_i)}{\sum_j \exp(x_j)}, \qquad (4.25)$$

where this activation function accesses all neurons of the output layer to calculate the sum in the fraction.

In the further course, we will get to know other activation functions. How do we choose the right activation function for a neural network or for the individual layers? There is no universally valid recipe. Many networks are optimized through tests and variations until a good mapping quality and low error have been achieved. Nevertheless, some rules can be set up that can help to some extent in developing the architecture of a network:

- If the output layer must reflect numerical values greater than 1, pure sigmoid or tanh functions are not suitable. In such cases, the use of ReLU often helps.
- For classification problems, the output layer should be activated with Softmax.

4.5.4 *Training of Neural Networks*

How in the case of the adaptive LMS algorithm, a neural network learns through the comparison of output and reference. We determine the error of the network by calculating an output value x_k with its current weights and thresholds and comparing this with the given target value d. Note that in the general case, the network output is multidimensional and thus also the training specification d. The training assumes that a training set with sufficient inputs x_0 and desired results d is available. Here too, we determine a quality functional J, to quantify the deviation of the network output x_k from the desired result d. A simple example of such a quality functional is (again) the square distance,

$$J = (x_k - d)^2. \tag{4.26}$$

There are other quality or cost functions that we could use for J. To explain the training of the network, however, we initially stick with this cost function. To train the network, we iterate over the training set. We systematically adjust the weights and thresholds so that J becomes smaller. If we minimize J, the network maps the input x_0 to $x_k \approx d$. The network learns the relationship between x_0 and d. To reduce J as efficiently as possible, we need to know how we can adjust the weights. The weights influence J in (4.26) via y,

$$J = \left[f\left(W_k x_k + b_y \right) - d \right]^2. \tag{4.27}$$

The derivative of the cost function with respect to the weights

$$\frac{\partial J}{\partial w_{ik}} = \frac{\partial J}{\partial f} \frac{\partial f}{\partial \mathrm{net}_k} \frac{\partial \mathrm{net}_k}{\partial w_{ik}}, \tag{4.28}$$

so the gradient of J with w_{ik}, points in the direction of the steepest increase of J with that very weight. As we have already seen in (4.10), we now only need to move in the direction of the steepest descent (i.e., the negative gradient) of J to deliberately minimize J. This is also referred to as the gradient descent method. We adjust each weight w_{ik} therefore via this gradient by changing it by

$$\Delta w_{ik} = -\eta \delta_i x_k \tag{4.29}$$

where δ_i is given by

$$\delta_i = \begin{cases} f_i'(\mathrm{net}_i)[x_i - d_i] & \text{if } i \text{ is the output layer} \\ f_i'(\mathrm{net}_i) \sum_j \delta_j w_{ij} & \text{else} \end{cases} \tag{4.30}$$

Specifically for the inner layers, we need to change the weights sequentially. We do this by starting from the output and moving forward through the network, changing the weights. This process is called **Backpropagation** and trains the network. For specific activation functions f, the change Δ_{ik} over (4.30) depends on the derivative f'.

4.5.5 Optimization as Hyperparameter

Already in our introduction, we mentioned the relationship between optimization and machine learning. The neural network optimizes the function J in such a way that it is minimized for a given set of training data. The derivation of Δw_{ik} in the previous section is illustrative. It explains to us how to analytically bring about the change in weights when J is defined as a square distance. The most important ingredient in (4.29) was the gradient of J from (4.28) and the gradient descent method. For gradient descent, optimized variants exist that we can use via the Python libraries:

- **Stochastic Gradient Descent** [14] (**engl.**). Whilethe gradient descent method uses the average of all available training data for its cost calculation, the SGD selects a subset of training data via a random generator. This new training set is smaller and the learning process is accelerated. The advantage of SGD is that this optimizer can also be used on very large data sets.
- **ADAM** [8], ADAGRAD [3] and **RMSProb** [12]. These three variations are based on the SGD. **ADAGRAD** varies the learning rate of the individual weights and leads to adaptive gradients. Frequently changed weights are assigned low learning rates and rarely changed weights are assigned high learning rates. This step is helpful to increase the convergence speed of the method. **ADAM** and **RMSProb** ensure that too high gradients have less influence and low gradients contribute more to the learning process.

4.5.6 Simple Classification Network with Scikit-Learn

We begin our implementation examples with a simple, pre-made neural network from the Scikit-Learn library. This collection of learning methods is excellent for getting to know the individual methods, experimenting with them, and setting up initial models.

But first, we load our data from the motor current example. Listing 4.6 shows how you can load and display this from the pickle file.

Listing 4.6 Loading and visualizing the data

```
import matplotlib.pyplot as plt
import pickle

data = pickle.load(open('EX03Engine.pickle','rb'))
X = data['X']

mycolor = []
for i in range(0,500):

    if data['Label'][i] == 1:
        mycolor.append('r')
    elif data['Label'][i] == 2:
        mycolor.append('k')
  '   elif data['Label'][i] == 3:
        mycolor.append('b')
    else:
        mycolor.append('y')

    plt.plot(X[i], color=mycolor[i],alpha=0.1)

plt.show()
```

After we have loaded the test data, we manually divide it into a training and a test set. In Listing 4.7 the arrays Xtrain, Ytest as well as Xtest and Ytest have been created for this purpose. Since we have a label in the dataset, the assignment here is particularly simple.

Listing 4.7 Splitting into training and test sets

```
Xtrain = []
Ytrain = []

Xtest = []
Ytest = []

for i in range(0,1000):
    Xtrain.append(X[i])
    Ytrain.append(data['Label'][i])

for i in range(1401,1420):
    Xtest.append(X[i])
    Ytest.append(data['Label'][i])
```

With the two data groups, we can now train a neural network. Our first step is to include a Multi-Layer Perceptron from the Scikit-Learn library. This can be included via `sklearn.neural_network.MPLClassifier` as Listing 4.8 shows. The call requires us to create an object `classifier`, to which we assign a corresponding `MLPClassifier` with special properties. In the present code example, the optimizer "lbfgs" is selected, the learning rate is initialized with 0.01 and in addition to an input and an output layer, a single hidden layer with 25 neurons is specified. The network is manageable and flat.

Listing 4.8 Neural network using Scikit-Learn

```
from sklearn.neural_network import MLPClassifier

classifier = MLPClassifier(
                  solver='lbfgs',
                  learning_rate_init=0.01,
                  hidden_layer_sizes=(25,)
                  )

classifier.fit(Xtrain, Ytrain)
```

After training the network, the immediate question is how good the model is. This is reflected in the cost function in the model, but we now want to test our model directly on the test data. For this, the function `predict` of the `MLPClassifier` class can be used. A simple comparison on the test set, as it is also helpful for other methods, is shown in Listing 4.9.

Listing 4.9 Test procedure for showing classification results

```
result = clf.predict(Xtest)

print('NN_|_Test')
for i in range(0,19):
    print('{}_|_{}'.format(result[i], Ytest[i]))
```

4.5.7 Classification Network in Keras

Individual Network Topology and Optimization

As simple as the implementation with the help of Scikit-Learn is, if you want to create more complex structures or even interact more deeply with the network, then the implementation with the help of the Tensorflow and Keras libraries is recommended. Keras gives us access to simple structures for programming models and Tensorflow contains fast, GPU and neural core optimized training algorithms.

Especially the previously discussed derivatives of activation functions and the optimization algorithms are included in Tensorflow. This leads to efficient and fast training processes. However, these advantages also mean that we have to set up a network in a more complicated way.

Categorical Mapping

One tool for developing a neural network for classification is categorical mapping. With a single neuron in the output, we can easily classify the states 0 and 1. But if there are N many, discrete categories, then we also need N neurons. Each neuron then stands for a class.

The categorical mapping is a rule

$$x \mapsto \sum_i \delta_{ix} e_i, \tag{4.31}$$

that assigns a unit vector e_i with the position x to a scalar classification number x.

In practice, the following assignment explains how this regulation works:

$$0 \mapsto [1,0,0,0,0],\ 1 \mapsto [0,1,0,0,0],\quad 2 \mapsto [0,0,1,0,0],$$
$$3 \mapsto [0,0,0,1,0],\ 4 \mapsto [0,0,0,0,1] \tag{4.32}$$

In Keras, there is a helper function that takes this step for us. This function is called `to_categorical()` and ensures that our categories are mapped to vectors with 0 and 1 accordingly.

Implementation of the classifier

Listing 4.10 shows a code example for classification with Keras. We need the model class from Keras. From this, our own Python object inherits all necessary properties and functions, here `class Classifier(Model)`. Note that we also need to initialize the model itself, so there is a corresponding call to the constructor of the superclass in line 9.

Listing 4.10 Classification network in Keras

```
import keras
import numpy as np
import tensorflow as tf
from keras.utils import to_categorical
from tensorflow.keras import layers, losses
from tensorflow.keras.models import Model


class Classifier(Model):

    def __init__(self, inputLayerLength, hiddenLayers=2):
        super(Classifier, self).__init__()
        self.inputLayerLength = inputLayerLength
        self.hiddenLayers = hiddenLayers
        self.constructLayers()
        self.classifier = tf.keras.Sequential(self.myLayers)
        self.compile(optimizer='adam',
                     loss=losses.CategoricalCrossentropy())
        self.optimizer.learning_rate = 0.001

    def constructLayers(self):
        self.myLayers = []
        self.myLayers.append(layers.Input(self.
            inputLayerLength))
        for i in range(0,self.hiddenLayers):
            self.myLayers.append(layers.Dense(50, activation='
                relu'))
        self.myLayers.append(layers.Dense(5, activation='
            softmax'))

    def call(self, x):
        classified = self.classifier(x)
        return classified

classifier = Classifier(len(Xtrain[0]))
history = classifier.fit(np.array(Xtrain),
                         to_categorical(np.array(Ytrain)),
                         epochs=50, batch_size=50)
```

By calling the function to_categorical(Ytrain), our categorization, as described above, is encoded on five neurons, which corresponds to the topology with 5 neurons in the output layer.

In the further course, this code example shows another feature: the construction of the layers is outsourced to an additional member function constructLayers of the class. Here you will find the input layer, the hidden layers, and the output layer. A ReLU function was specified as activation in the hidden layers. Since it is a classifier, a softmax activation function is used in the output layer. The chosen optimizer is ADAM and the cost function selected is the categorical cross-entropy. The initial learning rate is 0.001 in this case.

Epochs and Batch Size

For the call of the training, we must specify two parameters: the number of epochs and the batch size. An epoch stands for training on a dedicated data package, the batch, whose size is determined by the batch size. A small batch size thus means fast training of an epoch. For each epoch, the mix of data in the batch is randomly selected from the training data set Xtrain. Each epoch results in a final cost value for J.

Lines 30–33 finally specify this call. The training proceeds similarly to the example with Scikit-Learn. As a result, we get a classification model, which can be tested as in Listing 4.9. However, we need to vary our test because we have a prediction from 5 neurons whose activation tells us which category we have in the result. Listing 4.11 shows the test again, now with a modification in line 5: here the argmax of the classification result is now returned.

Listing 4.11 Test of the Keras classification network

```
result = classifier.predict(np.array(Xtest))

print('NN␣|␣Test')
for i in range(0,19):
    print('{}␣|␣{}'.format(np.argmax(result[i]), Ytest[i]))
```

Feel free to use this code to test different optimizers (SGD, RMSprob) and activation functions (e.g., ReLU or Leaky-ReLU).

Learning curve

In line 32 of Listing 4.10 we have stored the variable history. It contains all information about the learning process, including the course of costs in history.history['Loss']. This course is called the learning curve. In Fig. 4.6 the learning curve for the classifier training with the SGD optimizer is shown.

4.5.8 Regression network in Keras

Next, we want to train a regression network. For technical processes, regression is an established tool. Often, defined functions are used to set up parametric models with them. Polynomial regression is a prominent example of this.

To set up a simple initial model, we need a set of data. We want to keep this simple and therefore generate synthetic data ourselves. For this, we use a polynomial of the 2nd degree as a basic framework. The coefficients of the polynomial p fluctuate around a base value defined by us $p = (0.5, 1, 0.5)$.

Listing 4.12 shows the creation of the training and test set. In the function polynom(t,p), the actual polynomial is calculated. The function generateSyntheticData(t,n) generates n different polynomials each with the parameters p_i. The function plots represent our input data. They are stored in Xtrain

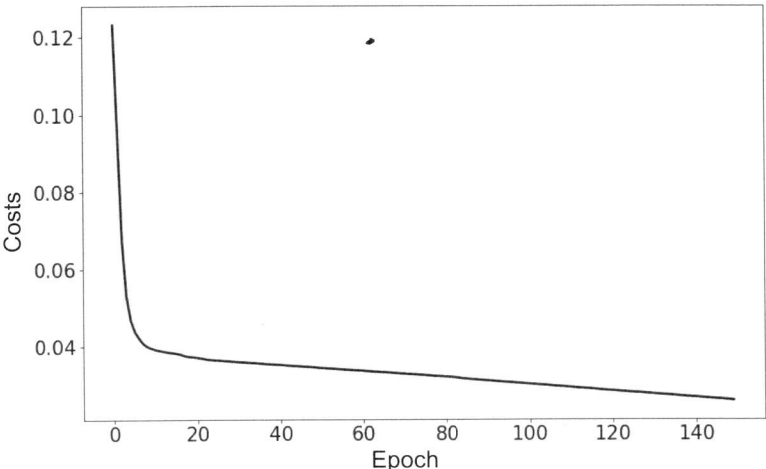

Fig. 4.6 Learning curve of the classification network

and `Xtest`. We save the strength of the quadratic component as our regression targets—i.e., our desired results—for both the training set `Ytrain` and the test set `Ytest`.

Listing 4.12 Synthetic data for a regression network in Keras

```
def polynom(t, p):
    f = 0
    for i in range(0,len(p)):
        f+=p[i]*t**i
    return f

def generateSyntheticData(t,n):
    result = []
    target = []
    p0 = [0.5, 1, 0.5]
    for i in range(0,n):
        p = p0+0.5*np.random.random(3)
        result.append(polynom(t,p))
        target.append(p[2])
    return result, target

Xtrain, Ytrain = generateSyntheticData(np.arange(0,5,0.2),500)
Xtest, Ytest = generateSyntheticData(np.arange(0,5,0.2), 12)

for i in range(0,len(Ytrain)):
    Xtrain[i] = Xtrain[i] - np.mean(Xtrain[i])

for i in range(0,len(Ytest)):
    Xtest[i] = Xtest[i] - np.mean(Xtest[i])
```

Please also note that we subtract the mean values from the function plots for both data sets. The training data generated in this way are now used as the basis for a neural regressor. This regressor is written very similarly to the above classifier. Listing 4.13 shows the code for this network.

Listing 4.13 Regression network in Keras

```
class Regressor(Model):

    def __init__(self, inputLayerLength, hiddenLayers=10):
        super(Regressor, self).__init__()
        self.inputLayerLength = inputLayerLength
        self.hiddenLayers = hiddenLayers
        self.constructLayers()
        self.regressor = tf.keras.Sequential(self.myLayers)
        self.compile(optimizer='adam',
                     loss=losses.MeanSquaredError())
        self.optimizer.learning_rate = 0.001

    def constructLayers(self):
        self.myLayers = []
        self.myLayers.append(layers.Input(self.
            inputLayerLength))
        for i in range(0,self.hiddenLayers):
            self.layers.append(layers.Dense(10, activation='
                relu'))
        self.myLayers.append(layers.Dense(1, activation='
            leaky_relu'))

    def call(self, x):
        regressorResult = self.regressor(x)
        return regressorResult
```

Please note the changes compared to the classifier. The cost function (English *loss*) has been adjusted and the mean square error has been used. We want to minimize this as much as possible during training. The categorical distance as in a classifier would not be able to do this. The output layer has only one neuron, which should take the value of the quadratic coefficient of our polynomial. As activation functions, we combine ReLU and Leaky-ReLU. The learning rate is set to 0.001. In addition, we use ADAM as the optimizer for minimizing the cost function.

With the command in Listing 4.14 we start the training.

Listing 4.14 Running the regression network

```
regressor = Regressor(len(Xtrain[0]))
history = regressor.fit(np.array(Xtrain), np.array(Ytrain),
    epochs=350, batch_size=100)
```

In a final step, we need to test the quality of the model. For this, we evaluate the regression network on the test inputs Xtest and compare the results of the model prediction with the true values Ytest, which exactly reflect our target parameters. A corresponding code for the test is given in Listing 4.15.

Listing 4.15 Plotting the results of the regression network

```
plt.plot(Ytest, 'k')
plt.scatter(range(0,len(results)), results, s=60, color='k')
```

The result of this test is shown in Fig. 4.7. Points indicate the actual expected values, with the fine connecting line merely helping the eyes to better follow the distribution of the points (as there are no data in between). The results of the regressor are marked with the symbol ×. The reconstruction of the quadratic part of the polynomial is extremely good in this case.

4.5.9   Overfitting and Cross-Validation

It can happen, that a learning process, not necessarily only neural networks, over-fits a dataset. This means that the process no longer trains only meaningful differences in the input data, but also learns the noise in the input data. However, this noise does not carry any additional information. Above all, it can no longer help to improve the learning results on the test data. In Fig. 4.8 (a) shows an ideal classification and (b) graphically illustrates the problem of overfitting. Due to the overfitting, the error during training decreases, leading us to the erroneous assumption that our prediction quality is constantly improving.

We can detect overfitting by comparing the training error with the error on the test data. Overfitted learning algorithms show a low error on the training data and a significantly higher one on the test data. For the application, this means that one trains, then looks at the error on both datasets, and as soon as the error on the training data significantly drops below the error on the test data, one is in a state of overfitting.

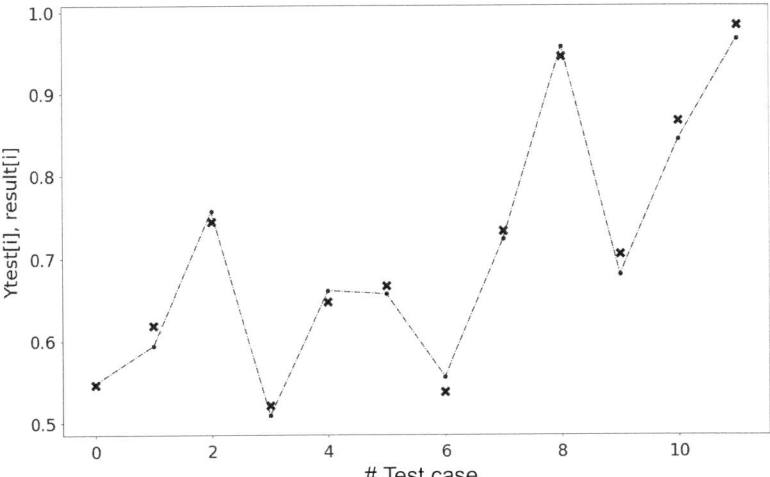

Fig. 4.7   Result of the regression network for predicting the quadratic part of a polynomial

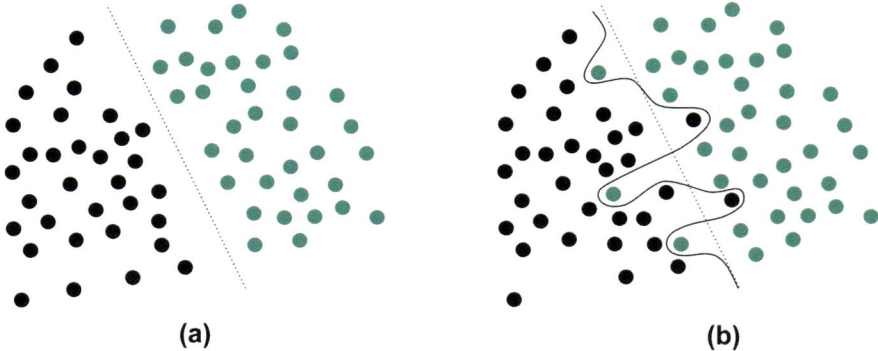

Fig. 4.8 Example of overfitting during the training of the separating surface of two classes. **a)** Idealized data without noise. **b)** Illustration of real data and the effect of overfitting

In section 4.2.3 we discussed the relevance of training and test data. The correct composition of these two sets determines the quality of the training. This is where the so-called cross-validation comes into play, which helps us to make ideal mixtures of data.

> We divide the set of all data into k equally large subsets
>
> $$T_1 = (X_1, Y_1), \ T_2 = (X_2, Y_2), \ \ldots, \ T_R = (X_k, Y_k), \qquad (4.33)$$
>
> We call these subsets **Folds.** Now we train the learning process k times. For this, one Fold T_i is defined, which is used for the test. The remaining $k - 1$ Folds T_m with $m \neq i$ are used for training. This procedure is called **Cross-validation.**

These folds can be quickly and easily generated with a helper function from Scikit-Learn. Listing 4.16 shows, based on already existing data X and labels Y, how to represent the above division process in Python.

Listing 4.16 Generate folds for a cross-validation

```
from sklearn.model_selection import KFold

kf = KFold(n_splits=10)

for i, j in kf.split(X):
    Xtrain, Xtest = X[i], X[j]
    Ytrain, Ytest = Y[i], Y[j]
```

You thus receive sets of Xtrain, Xtest, Ytrain and Ytest. You now divide your training and test runs onto these sets. The associated iteration of the fit call is part of our exercise task 4.7.

Fig. 4.9 Types of feedback
in recurrent networks. **a)**
Direct feedback, **b)** lateral
feedback, and **c)** indirect
feedback

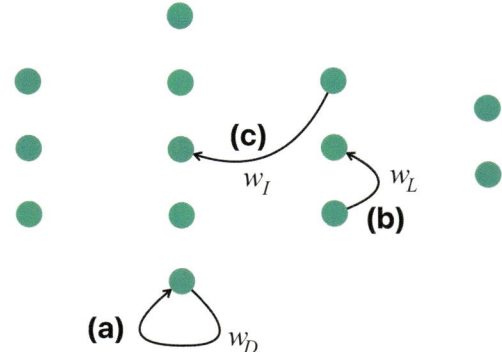

4.6   Recurrent Neural Networks

4.6.1   Feedback in Neural Network Structures

Recurrent neural networks deviate from the previous strategy of weight linkage.
They allow more complex linkages such as feedbacks of neurons. The following
new types of weight connections are distinguished (Fig. 4.9):

- **Direct feedback.** A weight vector is spanned from a neuron onto itself. That is,
  the output of the neuron is additionally fed back to its input, which generates
  the feedback. In Fig. 4.9, a recurrent network is shown. Case (a) corresponds to
  the direct feedback.
- **Lateral (side) feedback.** This leads the output of a neuron to the input of a
  neuron of the same layer. The flow of information in the network is thus per-
  pendicular to the evaluation direction. The lateral feedback is represented as
  case (b) in Fig. 4.9.
- **Indirect feedback.** Here the output of a neuron is fed back over one or more
  layers and connected with the input of a front neuron. This feedback is finally
  illustrated in diagram 4.9 as case (c).

In the next transformations, we want to illustrate the functioning of recurrent networks. To do this, we simplify our notation. We consider a single neuron with an
input x and an output y. Inside, this neuron has the (hidden) internal state h. As
with the adaptive filter, we use superscript, bracketed expressions to symbolize the
time steps. Due to the recurrence, the same neuron at time t can remember the
state at an earlier time step $t - 1$. The internal state $h^{(t)}$ is determined by the input
$x^{(t)}$, the weight w_{hx} for this input, the threshold b_h and the described previous value
for $h^{(t-1)}$,

$$h^{(t)} = f\left(w_{hh}h^{(t-1)} + w_{hx}x^{(t)} + b_h\right). \tag{4.34}$$

The output of the neuron is finally determined by

$$y^{(t)} = f\left(w_{yh}h^{(t)} + b_y\right), \tag{4.35}$$

where we use an output weight w_{yh} and another threshold b_y. Along the time axis, the feedback thus has an effect via its own weight w_{hh}.

Since this temporal dimension is of interest for feedback processes, it can be helpful to graphically represent the temporal flow through a neuron. To do this, we move from iteration steps at different times $(t-1)$, (t), $(t+1)$. In Fig. 4.10 we have illustrated this unfolding step.

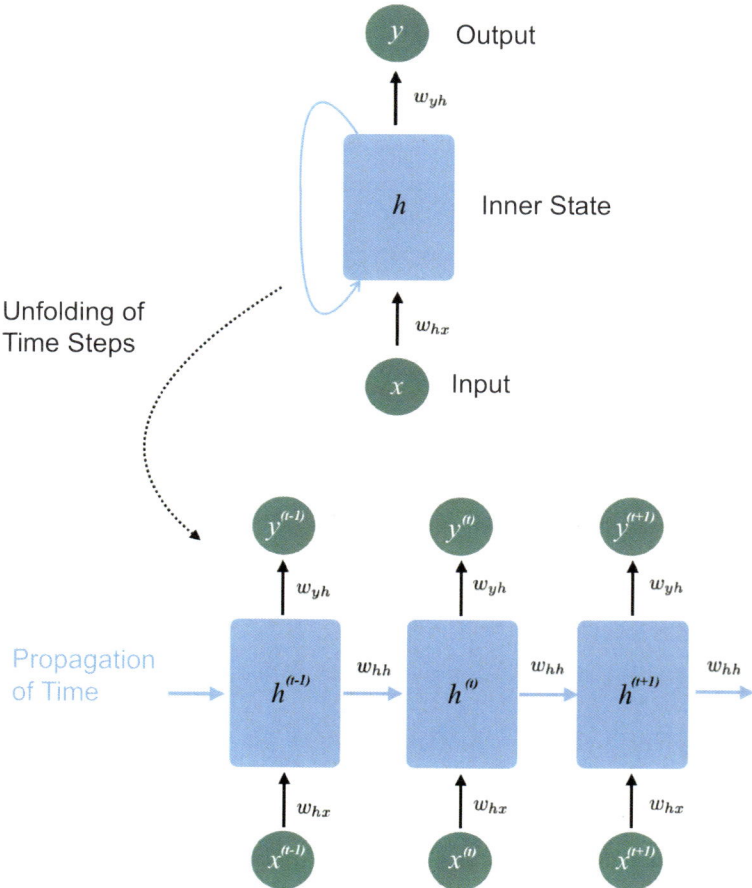

Fig. 4.10 Unfolding of the temporal iteration steps in the recurrent network

4.6.2   Gates for Controlling the Flow of Information within the Neuron

Feedback, like we just described, is a fundamental concept of control theory. This data fed back by the network is used in recurrent networks to specifically improve the learning process. However, we need a tool to control the data within a neuron. This tool is the gate. Like a valve, a gate can let data pass or block it. It does not act like a hard switch, but rather like a lock that has a variable opening.

In our case, we define the gate using another sigmoid function. The sigmoid maps its input to the value range between 0 and 1, thus realizing a suitable flow control. We abbreviate the gate with γ and use the index, here A, to refer it to a specific route,

$$\gamma_A(x) = \text{sig}(w_A x). \tag{4.36}$$

The shown gate has its own trainable weight w_A, which is multiplied with the gate input x and finally entered into the sigmoid function. γ_A thus acts as a dynamic weight factor. If this is applied to x_{k-1},

$$x_k = \gamma_A(x_{k-1}) x_{k-1}, \tag{4.37}$$

a large proportion of x_{k-1} reaches x_k, when γ_A is large, and less when γ_A is correspondingly small. The strength of γ_X thus controls what proportion x_{k_1} has on x_k. Without additional inflows, the single gate does not make much sense yet.

4.6.3   Neural Network with Long Short-Term Memory (LSTM)

Next, we want to give a neuron a memory m that lasts over several iterations. This idea goes back to Hochreiter and Schmidhuber [7] and in this case we are talking about a long short-term memory (LSTM). LSTM, along with alternative strategies (e.g., Gated Recurrent Units, GRU), has proven to be extremely successful for application to time series, as also shown by A. Graves in [5]. Speech recognition, automated parsing of texts, and image recognition problems are just a few examples where this technology is used. Therefore, we will specifically deal with the idea behind LSTM and show in the next section how we can use LSTM for our own networks.

The recurrent memory, which we will call m in the course of our considerations, we pass on, just like the hidden state h, from time step to time step. The gates introduced in the previous sections play a special role here. We define three different gates, a forget gate, an update gate, and an output gate. Fig. 4.11 shows a scheme of how the gates are arranged. The temporal change again runs from left to right, with h and m running in this direction. From bottom to top, the mapping from input x to output y is given.

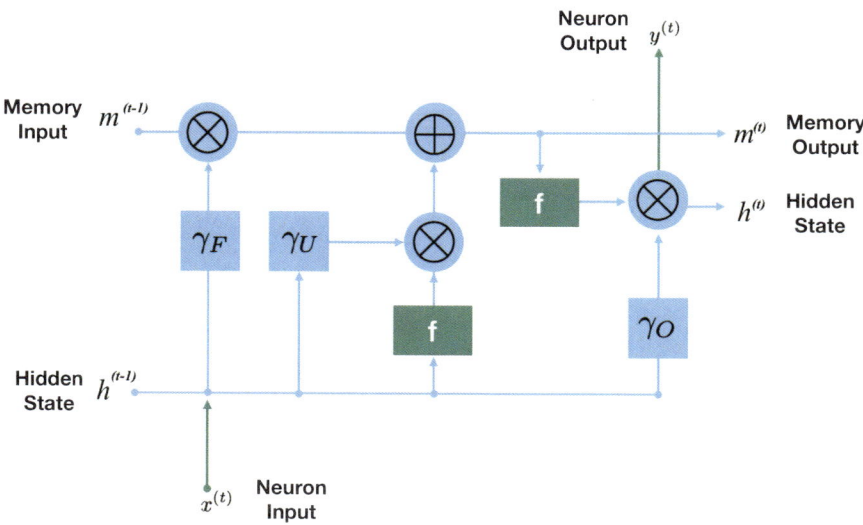

Fig. 4.11 Signal flow through a cell with long short-term memory (LSTM)

The Forget Gate γ_F is determined by

$$\gamma_F = \text{sig}\left(w_{Fh}h^{(t-1)} + w_{Fx}x^{(t)} + b_i\right) \qquad (4.38)$$

. It is capable of influencing the currently stored value in m. The Update Gate γ_U is given by

$$\gamma_U = \text{sig}\left(w_{Uh}h^{(t-1)} + w_{Ux}x^{(t)} + b_u\right) \qquad (4.39)$$

and the Output Gate γ_O is calculated by

$$\gamma_O = \text{sig}\left(w_{Oh}h^{(t-1)} + w_{Ox}x^{(t)} + b_o\right). \qquad (4.40)$$

From the hidden state $h^{(t-1)}$ and the input $x^{(t)}$ a new candidate for the memory can be determined—the potential update for m:

$$c_M = f_k\left(w_{hh}h^{(t-1)} + w_{hx}x^{(t)} + b_h\right). \qquad (4.41)$$

The new $m^{(t)}$ is determined from the candidate c_M and the old memory value $m^{(t-1)}$
,

$$m^{(t)} = \gamma_U c_M + \gamma_F m^{(t-1)}. \qquad (4.42)$$

One can imagine this equation as a mixture. Depending on the current pass-through factor of the gates, more of the candidate is added or not. The Forget Gate allows the memory of the previous memory value $m^{(t-1)}$ to be influenced. If γ_F is low, the memory value loses influence over time.

The network output is finally determined by

$$y^{(t)} = \gamma_O f_k(m^{(t)}),   \tag{4.43}$$

where m and h are iteratively passed on over time.

Please note that the above calculations apply to each neuron of a network. In networks with multiple layers, the steps must be performed for each individual neuron. The unfolding should not be confused with the network topology. The illustration in Fig. 4.10 applies to each individual neuron.

4.6.4   Implementation of an LSTM Network

The above equations contain a multitude of new weights, new thresholds, and influences of gates. Nevertheless, these new components of the network can be trained similarly using backpropagation. The complex step of modifying the weights in such a constellation again requires knowledge of the gradient of the cost function with respect to these specific weights. Once this is known, Δw can be determined to change the weights. These complex steps are performed using libraries and do not (fortunately) have to be implemented yourself.

In the following, we always assume networks that are fully recurrent. Libraries like Keras offer us the opportunity to use the capabilities of LSTM neurons for our networks and formulate them recurrently.

For our program example, we prepare training and test data in Listing 4.17. We again use the example file EXAMPLE02.pickle from Sect. 3.5.

Listing 4.17 Preparing example data for using LSTM

```
x = pickle.load(open('EXAMPLE02.pickle','rb'))

Xtrain = []
Ytrain = []
Xtest = []
Ytest = []
windowLength = 25
for i in range(0,5000-windowLength):
    Xtrain.append([x[i:i+windowLength]])
    Ytrain.append(x[i+windowLength+1])

Xtrain = np.array(Xtrain)
Xtrain = Xtrain.reshape(Xtrain.shape[0], 25,1)

for i in range(5001,5600-windowLength):
    Xtest.append([x[i:i+windowLength]])
    Ytest.append(x[i+windowLength+1])

Xtest = np.array(Xtest)
Xtest = Xtest.reshape(Xtest.shape[0], 25,1)
```

The aim of Listing 4.17 is to generate a training dataset that always takes a window of length `windowLength` from the variable `x` and finally takes the next following value in `x` as the target value for `Ytrain`. After that, corresponding windows are also built for `Xtest` and `Ytest`.

Listing 4.18 shows the code for the LSTM network. It imports a special layer for Keras that ensures recurrence.

Listing 4.18 Time series prediction using LSTM

```
import keras
import numpy as np
import tensorflow as tf
from tensorflow.keras import layers, losses
from tensorflow.keras.models import Model
from tensorflow.keras.layers import LSTM


class LSTMPredictor(Model):

    def __init__(self):
        super(LSTMPredictor, self).__init__()
        self.inputLayerLength = 1
        self.myLayers = []
        self.myLayers.append(layers.LSTM(25, input_shape=(1,
            25)))
        self.myLayers.append(layers.Dense(1, activation='relu'
            ))
        self.predictor = tf.keras.Sequential(self.myLayers)
        self.compile(optimizer='adam',
                    loss=losses.MeanSquaredError())
        self.optimizer.learning_rate = 0.005

    def call(self, x):
        predictor = self.predictor(x)
        return predictor
```

With the call in Listing 4.19 we train the network.

Listing 4.19 Running the prediction

```
lstmPredictor = LSTMPredictor()
history = lstmPredictor.fit(np.array(Xtrain), np.array(Ytrain), epochs
    =120, batch_size=30)
```

Listing 4.20 evaluates the network on the test data and thus predicts the following value for each new window.

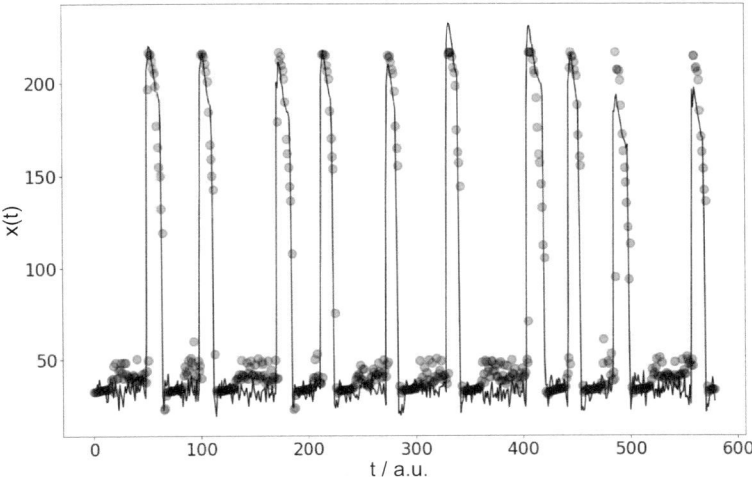

Fig. 4.12 Result of the LSTM prediction on the test data

Listing 4.20 Plotting the result of the time series prediction

```
results = lstmPredictor.predict(np.array(Xtest))

plt.plot(Ytest, 'k-')
plt.scatter(range(0,len(results)), results, s=110, marker='o',linewidth
    =2,color='k',alpha=0.3)
plt.tick_params('both', labelsize=22)
plt.xlabel('Testfall i', fontsize=24)
plt.ylabel('Ytest[i], result[i]', fontsize=24)
```

Fig. 4.12 shows the result of this code. The slightly transparent points represent the prediction of the network. The solid line shows us the true course. The LSTM network thus predicts the future course of the time series for us.

4.7 Decision Trees

4.7.1 Basic Idea of the Decision Tree

We now turn to a conceptually completely different learning method, the decision trees. We all know datasets where we can see at a glance which variables are related. Let's assume we have five columns of data and label them with x_{i1}, x_{i2}, x_{i3}, x_{i4} and x_{i5}. The index i here gives us the i-th data row in a dataset with N different measurements.

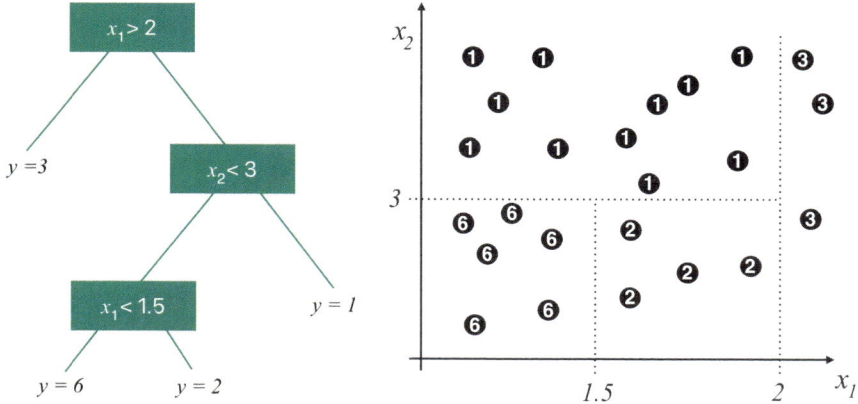

Fig. 4.13 Illustration of what a decision tree might look like (left) and a graphical representation of how the corresponding division in the attribute space X is realized (right). The numbers in the black dots stand for the y value of the target variable

We first define the following terms to be able to express our next steps more clearly in language:

> We call a variable y the **target variable** of a procedure if our intention is to be able to predict this variable with the help of the others. All data x_i that we use to predict y, we call **input data or attributes.** Here, $x_i \in X$ and $y \in Y$.

Without loss of generality, we assume the variable y to be a scalar value. Thus, reproducing the target variable is the task of this learning method. As before, we also distinguish here between classification and regression. The task of a decision tree is to successively determine the influence of each input variable on the target variable. It statistically determines the probability of the target variable carrying a certain result, depending on the configuration of the input data.

> A **decision tree** is a model $f_T : X \to Y$, which trains rules for the input data $x \in X$ to predict a target variable $y \in Y$.

Fig. 4.13 on the left shows an idealized example tree—here for a quite concrete scenario. The values are only for illustration. On the right side of Fig. 4.13 you can see the segmentation in the data space X, which here also consists of only two variables for illustration. The tree depicted here has correctly divided the various y values by defining rectangular areas in the X space.

Abstractly, an algorithm for such a tree uses a top-down method:

- **Splitting (Split).** Suitable criteria are found on the basis of which a division of the variables can be made. The most important variable, which contains the most information about the prediction variable, is at the beginning of the tree, as root.
- **Pruning (Pruning).** Large tree structures can memorize smallest details instead of correctly training the model. Therefore, the trees are kept as small as possible or pruned.
- **Recursive repetition.** The previous steps are repeated.

4.7.2   Entropy and Information Gain

Example: Branch Optimization
We consider a situation from the branch business. You operate several stores and sell goods. These goods come via a supply route (SupplyChain) and can be stored in a warehouse (Storage). You also have costs for the staff (StaffCost) and generate a certain turnover (Revenue). In Fig. 4.14 various example data are listed. All variables are captured by 0 or 1, which have the following meanings: The warehouse can exist (1) or not (0), the SupplyChain can function well (1) or not (0), the costs for salaries are either high (1) or low (0) and finally the turnover is high (1) or low (0). ◄

The question is first, what is the target variable in this example? Here, of course, the turnover is the obvious choice, because this is the size that we explicitly cannot influence ourselves. It is also the most sensible target size from the problem of optimization. All other variables, on the other hand, are controllable and input variables. The next question is, which of the variables has the strongest influence on the turnover. You can see this, in this selected example, already with the naked eye: the last column and the column for the warehouse (Storage) show a strikingly

	SupplyChain	Storage	StaffCost	Revenue
0	0	1	0	1
1	1	1	1	1
2	1	0	1	0
3	1	1	1	1
4	0	1	1	0
5	0	0	0	0
6	0	0	1	0
7	1	1	1	1
8	0	1	0	0
9	1	1	1	1

Fig. 4.14 Example data for optimizing business branches

high agreement. So one can put forward the thesis that whenever the warehouse is present, the turnover is also high.

This approach is very rudimentary. But it is already very close to the mathematical processes in the decision tree. Now if we count the favorable and possible cases, then there are turnovers with 1 in 6 cases. The table then contains 10 possible cases, so the probability is $6/10 = 1/2$ that a high turnover is present. There are also 7 warehouses. If we limit our focus only to the cases where the warehouse is present, we count all 6 cases with high turnover. Only in one of the 7 warehouse cases was the turnover low. The conditional probability for high turnover, under the condition that a warehouse exists, is therefore

$$P(\text{Revenue} == 1 \mid \text{Storage}==1) = \frac{6}{7} = 0.85, \tag{4.44}$$

which shows that this variable must have a high influence on the target variable. If a warehouse is present, there is a probability of 85 % that the turnover is high.

Entropy
Let's formalize this idea a bit more. We introduce the size entropy for this purpose.

The **entropy** H of a random variable x, is defined as

$$H(x) = -\sum_{x \in \Theta} P(x_i) \log_2 P(x_i) \tag{4.45}$$

It is a measure of the uncertainty of x. Θ is the event space of x.
Here, the sum runs over all possible states of x.

Example: Coin Toss
When we toss a coin, the possible states of $\Theta = \{\text{Head, Tails}\}$ and the probabilities $P(\text{Head}) = 0.5$ and $P(\text{Tails}) = 0.5$. The entropy then amounts to $H = -0.5 * \log_2(0.5) - 0.5 * \log_2(0.5) = 1$. ◄

A process that is completely random has maximum uncertainty (maximum disorder). The entropy in this case is $H = 1$. Low values of entropy thus mean a high certainty for the prediction, high values indicate a high randomness. In physics, entropy is also known as a measure of disorder.

For our specific example, we can now calculate the entropy for the turnover, here is $\Theta = \{1, 0\}$. We use the probabilities

$$P(\text{Revenue} = 0) = \frac{4}{10}$$
$$P(\text{Revenue} = 1) = \frac{6}{10} \tag{4.46}$$

and determine the entropy over (4.45) to

$$H_0(\text{Revenue}) = -\frac{6}{10}\log_2\left(\frac{6}{10}\right) - \frac{4}{10}\log_2\left(\frac{4}{10}\right) \approx 0.97 \qquad (4.47)$$

We call this value H_0, to indicate that it is a base value of entropy, without considering the other columns of the data. So if we only consider the turnover, the values are almost random. How can we now understand the influence of the other variables on our target size with the help of entropy? Are there variables (columns of the table) that exert an influence on the target column or by which we could predict the target variable?

For this, we limit the data to certain "what if" scenarios. We divide the above table for the variable StaffCost in such a way that two groups are formed, for StaffCost = 0 and StaffCost = 1. In Fig. 4.15 the rearranged tables can be seen. An ideal assignment would be given if our target variable in one half of the table, either top or bottom, only takes the same numerical values. This does not happen here, StaffCost does not directly predict the variable Revenue.

When you calculate the probabilities for the target variable, you determine conditional probabilities—conditioned by the state of the variables based on which you have divided the table. The entropy can also be calculated for this division based on the conditional probabilities and therefore one also speaks of the conditional entropy:

If x and A are two random processes, A can take the states a_i from the event space Θ_A. The **conditional entropy** is then defined as

$$H(x|A) = \sum_{a_i \in \Theta_A} P(a_i)H(x|A = a_i). \qquad (4.48)$$

	SupplyChain	Storage	StaffCost	Revenue
0	0	1	0	1
5	0	0	0	0
8	0	1	0	0

	SupplyChain	Storage	StaffCost	Revenue
1	1	1	1	1
2	1	0	1	0
3	1	1	1	1
4	0	0	1	0
6	0	1	1	1
7	1	1	1	1
9	1	1	1	1

Fig. 4.15 Division of data rows for low and high StaffCost values

In a clear way, the conditional entropy thus goes through all states of the variable A. It uses the entropy in each subgroup of the data and weights each state with its probability of occurrence $p(a_i)$. We test this again in our example above. First, we consider the category "StaffCost". The conditional probabilities for this can be read off in Fig. 4.15,

$$P(\text{Revenue} = 1|\ \text{StaffCost} = 0) = \frac{1}{3},$$
$$P(\text{Revenue} = 0|\ \text{StaffCost} = 0) = \frac{2}{3},$$
(4.49)

which ultimately leads to an entropy

$$H(\text{Revenue}|\ \text{StaffCost} = 0) = -\frac{1}{3}\log_2\left(\frac{1}{3}\right) - \frac{2}{3}\log_2\left(\frac{2}{3}\right) \approx 0.91. \ (4.50)$$

. Similarly, the entropy for the second half of the table is found,

$$H(\text{Revenue}|\ \text{StaffCost} = 1) = -\frac{5}{7}\log_2\left(\frac{5}{7}\right) - \frac{2}{7}\log_2\left(\frac{2}{7}\right) \approx 0.86. \ (4.51)$$

Both are the entropies in the summands of (4.48). The weights $P(a_i)$ in 4.48 are the probabilities that StaffCost takes either the value 1 or 0, and we can count these just as easily. The total conditional entropy $H(\text{Revenue}|\text{StaffCost})$ is thus (Fig. 4.15)

$$H(\text{Revenue}|\ \text{StaffCost}) = \frac{3}{10}*0.91 + \frac{7}{10}*0.86 = 0.875. \qquad (4.52)$$

The StaffCost column thus has a lower entropy for determining the target variable Revenue than if we had only looked at the target variable. Now let's look at the variable Storage, which we have already identified as extremely influential, resulting in a conditional entropy of

$$H(\text{Revenue}|\ \text{Storage}) = \frac{3}{10}*0.0 + \frac{7}{10}*0.59 = 0.413, \qquad (4.53)$$

even lower than for StaffCost. The information gain through the column is therefore higher than through StaffCost. We can therefore determine the influence of a variable on the prediction of our target size by the conditional entropy.

Information Gain
If we reduce the entropy, we gain information. Exactly this relationship can be described by the following new size (Fig. 4.16):

We define the **information gain** (Information Gain) IG by the variable A, as the distance between the original entropy $H_0(x)$ for the prediction of a target variable x, and the conditional entropy $H(x|A)$,

Fig. 4.16 First node of a decision tree

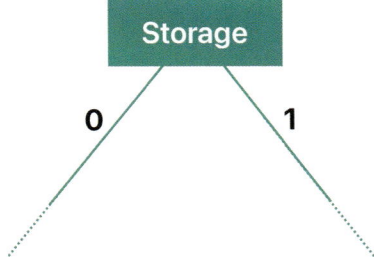

$$\text{IG}(x, A) = H_0(x) - H(x|A). \tag{4.54}$$

The **Information Gain** (IG) gives us the distance from the base entropy of a target variable to the conditional entropy by restricting to an input variable. The greater this distance, the more information the corresponding column has contributed to our criterion. For the above example, the following values result

$$\text{IG}(\text{Revenue}|\text{StaffCost}) = 0.97 - 0.875 \approx 0.095,$$
$$\text{IG}(\text{Revenue}|\text{Storage}) = 0.97 - 0.413 \approx 0.56, \tag{4.55}$$
$$\text{IG}(\text{Revenue}|\text{SupplyChain}) \approx 0.12.$$

This result confirms the impression that Storage has the most influence on the result. The first node is therefore Storage, see Fig. 4.16. Here the information gain is greatest. Accordingly, the tree is built up iteratively, as Fig. 4.17 shows.

Construction of a simple decision tree by division

To create a decision tree, we now follow an algorithm of the form:

- **(1) Calculate IG.** In a dataset, the IG for all input variables with respect to a target variable is calculated.
- **(2) Leaf.** The input variable with the highest IG is set as a node of a decision tree. The influence of this size is thus processed.
- **(3) Branching.** Through the tree root, two new datasets are created that correspond to the division based on the input variables.
- **(4) Iteration.** For each new data table, we iteratively return to step (1) and continue the process until all variables are processed.

The (simplified) algorithm shown here goes back to R. Quinlan [11] and is called Iterative Dichotomyzer (ID). In fact, we only consider the process of division. The algorithms C4.5 and C5.0 (both decision trees) also use a division over the IG. It is perhaps the most illustrative approach to understanding the construction of decision trees. There is no restriction on the number of splits, which unfortunately can lead to broad tree structures. Our example was chosen in such a way that this effect cannot occur here.

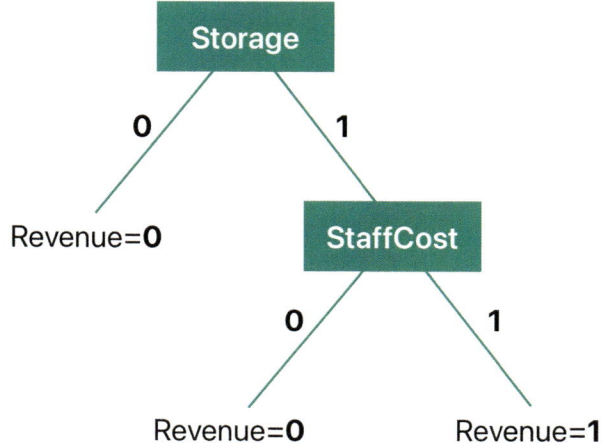

Fig. 4.17 First node of a decision tree

4.7.3 Classification and Regression Trees

The branching into many branches from a node is a disadvantage of the above
method, as it promotes wide tree structures. L. Breiman et al. showed in [2] an
approach based on binary divisions: the Classification-and-Regression-Tree algorithm (CART). Each node can only develop two branches here.

To specifically train a decision tree, a cost function J is again needed. It
describes the deviation of the tree output y to the desired label d (be it a classification or a regression),

$$J = \sum_{i=0}^{N-1} \left[f_T(x_i) - d_i \right]^2 \tag{4.56}$$

starting from a dataset with N data rows x_i and N labels y_i. This cost function is
already well known to us, it is again the mean square distance.

To counteract any growth of the tree, a penalty term can be added to the cost
function. This term is supposed to increase the costs if the tree becomes too large.
If K is the number of tree branches, then we choose this factor as $R = \alpha K$.

In summary, we can write the training of a decision tree as:

A decision tree is trained via an algorithm that divides the space X into K
disjoint subsets $\mathcal{X}_k$,

$$X = \mathcal{X}_0 \cup \mathcal{X}_1 \cup \cdots \cup \mathcal{X}_K, \tag{4.57}$$

so that over

$$\text{minimize} \qquad J = \sum_{i=0}^{N-1} (f_T(\boldsymbol{x}_i) - d_i)^2 + \alpha K \qquad (4.58)$$

$$\text{u. d. B.} \quad x_{ik} \in \mathcal{X}_k \wedge \mathcal{X}_k \cap \mathcal{X}_m = \emptyset \text{ für } k \neq m,$$

the cost function J is minimized.

Each vector $\boldsymbol{x}_i$ thus consists of k entries. Through the arrangement of the $\mathcal{X}_k$ of the tree, this minimization ultimately results in a model with $y = f_T(\boldsymbol{x}) \approx d$.

4.7.4 Application Example with Scikit-Learn

We now want to deal with the application of such a decision tree. In Python, there are several libraries that support us in training such tree structures.

Listing 4.21 Decision tree classifier with Scikit-Learn

```
%matplotlib tk
import matplotlib
import matplotlib.pyplot as plt
from sklearn.tree import DecisionTreeClassifier
from sklearn import tree
import numpy as np
from scipy import stats

import pandas as pd

company = pd.DataFrame()
company['SupplyChain'] =[0,1,1,1,0,0,0,1,0,1]
company['Storage']   =  [1,1,0,1,0,0,1,1,1,1]
company['StaffCost'] =  [0,1,1,1,1,0,1,1,0,1]
company['Revenue']   =  [1,1,0,1,0,0,1,1,0,1]
cpm = pd.DataFrame(company)

X = []
for i in range(0,len(company['Storage'])):
    x = [company['StaffCost'][i],company['SupplyChain'][i],
        company['Storage'][i]]
    X.append(x)
Y = company['Revenue']

clf = DecisionTreeClassifier(criterion='entropy')
clf.fit(X, Y)
fig = plt.figure(figsize=(12,4),dpi=100)
tree.plot_tree(clf)
```

Listing 4.21 shows the application of the decision tree from the Scikit-Learn library. The tree can be graphically output and interpreted using the function `tree.plot_tree`.

4.7.5   Implementation of a Simple Decision Tree

We now want to program a very simple tree, which you are free to expand as you wish. We only use Numpy as a base for this. Abstractly, we start with an "object shell", as listed in Listing 4.22. It starts with a class called simpleTree. We follow the basic ideas laid out in [10] by Quinlan, but simplify the individual steps for the sake of illustrating the methodology.

After defining a structure of split, fit and predict calls, we define placeholders for the depth of the tree, self.treeDepth, and the maximum branching depth self.treeDepthMaximum in the constructor __init__. They will be systematically expanded later with the code for training and prediction.

Listing 4.22 Simple, unfinished structure of a decision tree

```
import numpy as np

class simpleTree(object):

    def __init__(self, maxDepth:int):
        self.treeDepth = 0
        self.treeDepthMaximum = maxDepth

    def split(self, xtrain:np.array, ytrain:np.array):
        pass # muss noch geschrieben werden

    def fit(self, xtrain:np.array, ytrain:np.array, node={},
        depth:int=0):
        pass # muss noch geschrieben werden

    def predict(self, xtest:np.array):
        pass # muss noch geschrieben werden
```

The first ability our decision tree must have is the calculation of entropy. Since we (similar to the CART algorithm) only want to consider binary splits, we calculate the entropy for one branch with A elements and another branch with B elements. Both are int variables. The total number of all elements is $C = A + B$. The entropy tells us how good this split attempt was. Listing 4.23 adds this calculation to the tree. Here, only formula (4.45) is applied.

Listing 4.23 Binary entropy

```
    def binaryEntropy(self, A:int, B:int):
        if A == 0 or B == 0:
            return 0
        else:
            return  -(A*1.0/(A+B))*np.log2(A*1.0/(A+B)) \
                    -(B*1.0/(A+B))*np.log2(B*1.0/(A+B))
```

The binary entropy is needed to calculate the balance for the division of the set ytrain. We divide this set using a boolean array *b* into a "left" part and a "right" part. The left branch thus takes the *y* values where the value *b* was True, the right branch takes the remaining set. In Listing 4.24, this part of the calculation is shown and a total entropy value *H* for a specific vector *b* is determined.

Listing 4.24 Distribution using binary entropy

```
 1    def H(self, b:bool, ytrain:np.array):
 2        left = ytrain[z]
 3        right = ytrain[~z]
 4        nLeft, hLeft = 0,0
 5        nRight, hRight = 0,0
 6        for yd in set(left):
 7            nLeft = sum(left==yd)
 8            hLeft += float(nLeft)/self.N * self.binaryEntropy(
                  sum(left==yd), sum(left!=yd))
 9        H = float(nLeft)/self.N * hLeft
10        for yd in set(right):
11            nRight = sum(right==yd)
12            hRight += float(nRight)/self.N * self.
                  binaryEntropy(sum(right==yd), sum(right!=yd))
13        H += float(nRight)/self.N * hRight
14        return H
```

With the help of these two subfunctions, we can perform the split of the training and target variables in Listing 4.25. This split works in such a way that we iterate through each *x* value and use *b* to minimize the entropy. The cutoff is the *x* value at which we divide the branches as optimally as possible.

Listing 4.25 Performing the branch split with binary entropy

```
 1    def split(self, xtrain:np.array, ytrain:np.array):
 2        column = None
 3        cutoff = None
 4        entropyMinimum = 100
 5        localEntropyMinimum = 1
 6        for i, xc in enumerate(xtrain.T):
 7            for value in set(xc):
 8                b:bool = xc < value
 9                theEntropy = self.H(b, ytrain)
10                if theEntropy <= localEntropyMinimum:
11                    localEntropyMinimum = theEntropy
12                    theCutoff = np.round(float(value),4)
13
14            if theEntropy == 0:
15                return i, theCutoff, theEntropy
16
17            elif theEntropy <= entropyMinimum:
18                entropyMinimum = theEntropy
19                column = i
20                cutoff = np.round(float(theCutoff),4)
21
22        return column, cutoff, entropyMinimum
```

While `split` calculates a division for a specific dataset of the training set, the function `fit` in Listing 4.26 actually goes through all the data of the training set. This is where the actual tree is created. In line 12, the above `split` is called. Afterwards, the data is divided according to the results determined here for the `cutoff`. Lines 15–20 show the creation of each node of the tree. Here, we use the median of the *y* values for the calculation of the leaf value.

The most important step is finally the recursion for the left and right branch, where we again call `fit`. The process is thus continued in a nested manner and only stopped when the specified maximum tree depth has been reached.

Listing 4.26 Fitting the data into the tree

```
def fit(self, xtrain:np.array, ytrain:np.array, node={},
        depth:int=0):
    self.N = len(ytrain)
    if node is None:
        return None
    elif np.shape(ytrain)[0] == 0:
        return None
    elif all(x == ytrain[0] for x in ytrain):
        return {'Leaf': np.round( float(ytrain[0]), 4) }
    elif depth >= self.treeDepthMaximum:
        return None
    else:
        col, cutoff, entropy = self.split(xtrain, ytrain)
        ytrain1 = ytrain[xtrain[:, col] < cutoff]
        ytrain2 = ytrain[xtrain[:, col] >= cutoff]
        node = {
                'Index':col,
                'Cutoff':cutoff,
                'Leaf': np.round(np.median(ytrain))
                }
        node['Left'] = self.fit(xtrain[xtrain[:, col] <
            cutoff], ytrain1, {}, depth + 1)
        node['Right'] = self.fit(xtrain[xtrain[:, col] >=
            cutoff], ytrain2, {}, depth + 1)
        self.treeDepth += 1
        self.tree = node
        return node
```

The tree described here can already be trained with the given functions. However, it cannot yet be tested, as we have not yet created access to meaningfully read the tree again. In Listing 4.27, we therefore supplement the function `predict`. It takes test data `xtest` and goes through the tree according to its branching. To do this, we iterate over each data vector that is present in the set `xtest`. Afterwards, we use the *x* value to iterate to the corresponding position of the tree and find the last leaf that contains the target value.

Listing 4.27 Using the tree to predict results

```
def predict(self, xtest:np.array):
    results = []
    for eachTest in xtest:
        classResult = 0
        tree = self.tree
        check = True
        lastLeaf = 0
        while check:
            if eachTest[tree['Index']] < tree['Cutoff']:
                tree = tree['Left']
            else:
                tree = tree['Right']
            if tree!=None:
                if 'Leaf' in tree:
                    lastLeaf = tree['Leaf']
                if 'Index' in tree:
                    check = True
                else:
                    check = False
            else:
                check = False
        else:
            results.append(lastLeaf)
    return results
```

If you put the tree together into an object `simpleTree()`, we can train this tree with the code in Listing 4.28. Here we again use our established motor current example, which includes a classification.

Listing 4.28 Calling the simple tree and applying it to an example

```
import pickle
data = pickle.load(open('EX03Engine.pickle','rb'))
X = data['X']
Y = data['Label']
clf = simpleTree(maxDepth=10)
clf.fit(np.array(X[0:80]),np.array(Y[0:80]))
```

To perform the test, we use a range of data that we have not used for training and revert to our simple verification of the results. Listing 4.29 shows how the test is called.

Listing 4.29 Test of the simple tree

```
result = clf.predict(np.array(X[100:120]))
for i in range(0,19):
    print('{} | {}'.format(np.round(result[i],2), Y[100+i]))
```

The expected test result of our simple tree leads to a good classification, as the output in 4.30 illustrates.

Listing 4.30 Result of simple tree

```
1 | 1
2 | 2
3 | 3
2 | 2
1 | 1
2 | 2
2 | 2
2 | 2
2 | 2
```

The simple tree shown in the previous section can serve as a starting point for your own extensions. It illustrates the basic concept and explains the division of the branches. The core element of the division was the calculation of entropy. Alternatively, you could certainly also maximize the Information Gain (IG), which leads to equivalent results.

4.7.6 The Gini Coefficient for Decision Trees

The Gini coefficient, named after Corrado Gini [4], is a measure of concentration for sets. It is, alternatively to entropy, a popular tool for determining the quality of mixtures. The Gini index, as it is also called, indicates the deviation of data points from a uniform distribution. It therefore offers another way to determine the quality of a division.

The **Gini coefficient** is defined as

$$G(a) = \sum_{a_i \in \Theta_A} p(a_i)(1 - p(a_i)). \tag{4.59}$$

and indicates how much the values x_i deviate from x a uniform distribution.

In the code in Listing 4.31, we calculate the Gini coefficient for a binary division. This will later allow us to use the code in our above decision tree.

Listing 4.31 Gini coefficient

```
def gini(x):
    a = np.sum(x==True)
    b = np.sum(x==False)
    pa = float(a)/(a+b)
    pb = float(b)/(a+b)
    g = pa*(1-pa) + pb*(1-pb)
    return g
```

However, we first want to experiment in Listing 4.32 to see how the Gini coefficient behaves.

Listing 4.32 Understanding the Gini coefficient

```
x = []
x.append(np.array([True, True, True, True, True, True]))
x.append(np.array([False, True, True, True, True, True]))
x.append(np.array([False, False, True, True, True, True]))
x.append(np.array([False, False, False, True, True, True]))
x.append(np.array([False, False, False, False, True, True]))
x.append(np.array([False, False, False, False, False, True]))
x.append(np.array([False, False, False, False, False, False]))

for eachX in x:
    print(gini(eachX))
```

The output of these examples yields the values in Listing 4.33.

Listing 4.33 Result of Listing 4.32

```
0.0
0.2777777777777778
0.4444444444444445
0.5
0.4444444444444445
0.2777777777777778
0.0
```

The Gini coefficient in this implementation takes values between 0 and 0.5. To use it instead of entropy in the decision tree, we integrate the following listing:

Listing 4.34 Binary Gini coefficient for use in our tree

```
    def binaryGini(self, A:int, B:int):
        pa = float(A)/(A+B)
        pb = float(B)/(A+B)
        g = pa*(1-pa) + pb*(1-pb)
        return g
```

This function has the same call as the `binaryEntropy` function, so the tree can easily be switched to the alternative measure: You now only need to replace the call of the `xtest` with `binaryGini`.

Summary

In this chapter, we first dealt with the topic of learning in general and finally defined computer-aided learning as the result of an optimization process. We get to know two significant strategies in this book: supervised and unsupervised learning. In the course of this chapter, we focused on supervised learning, which learns a function f with $y = f(x)$, where data for x and y are available. The y are the labels or the target values we train on, and x are the input data.

The evolutionary algorithm learns by pure trial and error. It knows, at least in the variant shown here, no better strategy than to vary and test. This strategy is very similar to that of the game. As simple as this approach appears, this algorithm has already been successfully used in technical applications. Therefore, it was important for us to present this procedure here.

The LMS algorithm is a fundamental concept in control theory. It changes its filter parameters and thus learns to achieve a target size. Due to its simple structure, it is well suited as a first example of learning methods. Its basic training is very similar to that of the neural network.

Then we introduced neural networks. They play a prominent role in today's machine learning. Neural networks train weight and bias values in such a way that the network performs the correct mapping from input to output data.

Finally, we discussed decision trees. Here we first learned concepts for building the trees: entropy and Gini coefficient. Then we discussed both the Scikit-Learn decision tree and programmed our own Hello-World decision tree.

After this chapter, you should be familiar with the basic terms for learning methods. The aim here is not to present the full range of all possible methods, but to discuss the most important ideas and concepts using clear programming examples.

Tasks

4.1 Insert a counter into the code of the evolutionary algorithm that records the number of iterations in the `while` loop. Then add another counter to record the number of successful change processes. What is the ratio of the two numbers when you calculate multiple runs?

4.2 Test various activation functions in the implementation of the regression network: a) Sigmoid, b) Tanh, and c) ReLU. What happens to the learning curve? What influence does the choice of activation function have on the quality, in relation to this specific example?

4.3 Apply the Scikit-Learn implementation of the decision tree to our problem with the motor currents. What difference do you notice, by switching from Gini coefficients to an entropy-based distinction?

4.4 How could you integrate the quality function J from (4.58) into our simple decision tree `simpleTree`? At which points would you need to change the code?

4.5 Revisit the classification network with Keras from 4.5.6 and plot the learning curves for the same network, but with a) ADAM optimizer, b) RMSProb optimizer, and c) ADAGRAD optimizer.

4.6 Plot the learning curve for the regression network in Keras from 4.5.8.

4.7 Modify the training and test data for the LSTM network in Sect. 4.6.4 to model the motor current example data. Note: First, string all data vectors together to form a single, long time series. Then use the split demonstrated in Sect. 4.6.4 before you train the network.

4.8 Train the regression network from 4.5.8 using cross-validation! To do this, write the call to training (`fit()`) in a loop and successively evaluate the test results. Vary the test results between the folds.

4.9 Integrate the Gini coefficient into the `simpleTree`. To do this, replace the call to `binaryEntropy` in Listing 4.24 with the call to the function `binary-Gini` from 4.34.

References

1. C. M. Bishop, *Neural Networks for Pattern Recognition*. Oxford University Press, 1996.
2. L. Breiman, J. H. Friedman, R. A. Olshen, and C. J. Stone, *Classification and Regression Trees*. Taylor and Francis, 1984.
3. J. Duchi, E. Hazan, and Y. Singer, „Adaptive subgradient methods for online learning and stochastic optimization," *Journal of Machine Learning Research*, vol. 12, no. Jul, pp. 2121–2159, 2011.
4. C. Gini, „Measurement of inequality of incomes," *The Economic Journal*, vol. 31, no. 121, pp. 124–, 1921.
5. A. Graves, „Generating sequences with recurrent neural networks," 2014.
6. S. Haykin, *Adaptive Filter Theory*. Prentice-Hall, 2002.
7. S. Hochreiter and J. Schmidhuber, „Long short-term memory," *Neural computation*, vol. 9, no. 8, pp. 1735–1780, 1997.
8. D. Kingma and J. L. Ba, „Adam: A method for stochastic optimization," in *International Conference on Learning Representations*. International Conference on Learning Representations, 2015.
9. S. Koike, „A new efficient method of convergence calculation for adaptive filters using the sign algorithm with digital data inputs," *Acoustics, Speech, and Signal Processing, IEEE International Conference on*, vol. 3, p. 2333, 1997.
10. Tinterwordspacing J. Quinlan, „Simplifying decision trees," *International Journal of Man-Machine Studies*, vol. 27, no. 3, pp. 221–234, 1987. [Online]. Available: https://www.sciencedirect.com/science/article/pii/S0020737387800536
11. R. Quinlan, *Learning efficient classification procedures*.Springer, 1983.
12. M. Riedmiller and H. Braun, „A direct adaptive method for faster backpropagation learning: the rprop algorithm," in *IEEE International Conference on Neural Networks*, 1993, pp. 586–591 vol.1.
13. B. D. Ripley, *Pattern Recognition and Neural Networks*, 1st ed. Cambridge University Press, 2008.
14. S. Ruder, „An overview of gradient descent optimization algorithms," *arXiv preprint*arXiv:1609.04747, 2016.
15. B. Widrow and M. Kamenetsky, *Least-Mean-Square Adaptive Filters*, S. Haykin and B. Widrow, Eds. Wiley, 2003.

Chapter 5
Unsupervised Learning

Keywords Unsupervised learning · Principal component analysis · K-Means clustering · Autoencoder · Anomaly detection

Unsupervised learning seeks to uncover structures in the data. Unlike supervised learning, there are no labels or target sizes in unsupervised learning. Only the data itself is the starting point of an unsupervised learning process and therefore these methods are excellently suited for preprocessing. Even more: through their special analysis of the data, some of the methods find simpler, low-dimensional representations of the input data. They thus reduce the dimension and enable a better view of the relevant information. Our focus in this chapter will be on the classic Principal Component Analysis (PCA), the K-Means clustering method, the t-Distributed-Stochastic-Neighbour-Embedding algorithm (t-SNE) and the concept of autoencoders.

Unsupervised learning analyzes the structure in data. In doing so, it can extract important information for supervised learning: the labels or target sizes. Even more important is its ability to learn efficient transformations, such as the PCA or other dimension-reduced representations. This leads to a new form of data preprocessing. Chapter 5 links back to Chap. 3 and indirectly provides optimal input data for supervised learning methods.

Fig. 5.1 provides an overview of the impact of Chap. 5 on the other sections of the book. It also influences explainability, as understanding the structure of input data improves the explainability of models.

M. J. Neuer, *Machine Learning for Engineers*,
https://doi.org/10.1007/978-3-662-69995-9_5

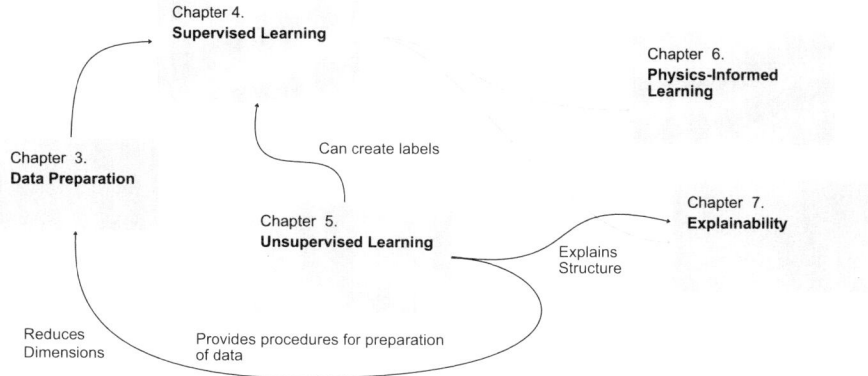

Fig. 5.1 Overview of the connection of Chap. 4 with the following chapters

5.1   Unsupervised Learning Paradigms

Creating models for variables is an obvious strength of supervised learning. It is based on the provision of input data and desired output behavior, input and labels, so that a model learns the relationship between the two. In many cases, however, we do not have labels for the training. Imagine a dataset where you have not yet selected any target quantity. Or think of a situation where you have collected process data but have no information about whether the products produced are faulty or not.

In all these cases, there is no specification of a label and no target variable for a regression. However, there are learning methods that can say something about the data even in this situation. They systematically establish relationships between the data points and try to identify an underlying structure—if there is one. In some cases, these approaches are able to reduce data in their dimension. The latter is also an idea that runs through the preprocessing steps from Chap. 3. Please note, all unsupervised learning methods can also be used for preprocessing for supervised learning. The following methods belong to the group of unsupervised learning and will be in the focus of this chapter:.

- **Principal Component Analysis.** Perhaps the best-known method for preprocessing data is the Principal Component Analysis. It uses a suitable transformation, the Karhunen-Loevé transformation, to order the importance of the respective eigenvectors in the eigen space of the covariance matrix and to specifically delete them. This creates a low-dimensional subspace that adequately represents our data.
- **K-Means Clustering Method.** A possible structure formation in data are clusters, i.e., data points that lie close together in larger groups. Determining such clusters is important as it helps us to find out to what extent different regimes

are represented in data. The K-Means method iteratively searches for the centers of such clusters.

- **t-Student Stochastic Neighbour Embedding.** This method includes two important aspects. It uses the distribution and the neighborhood relationships of data points in high-dimensional spaces to finally map them onto a low-dimensional space with a t-student distribution. Thus, a map of the data is found in which nearby points lie together in clusters.
- **Autoencoders.** The autoencoder is a neural network. Due to its individual topology, it only needs input data for its training and is therefore an unsupervised learning method. Autoencoders are capable of learning data sets, condensing them, and recognizing them again. They are used for both dimension reduction and anomaly detection.

5.2 Principal Component Analysis (PCA)

5.2.1 Properties of the Covariance Matrix

In Chap. 2 we have already learned about the covariance matrix. What kind of information is contained in this matrix? On the one hand, it tells us how much data columns or rows differ from each other. If the data points are far apart, we say the dispersion of the points is very high and so is the covariance. If points are close together and show little variance, the covariance is very low. In these cases, we are explicitly interested in the absolute position of the data points and do not use the correlation matrix, which would already be scaled by its definition.

The covariance matrix, by fully reflecting the statistical deviations of the data series from each other, also contains the information about which of the dimensions in the data play a role at all:

> **Example: Marbles in a Line**
> Imagine positioning marbles along an exact line on a table. You ensure that the marbles have as different distances from each other as possible, but never leave the course of the line. Let's also call the dimension along the line X and the orthogonal axis Y, on which, as already mentioned, there should be as little deviation from the line as possible. ◄

The covariance of the data then only occurs in X and not in Y. Is the consideration of the dimension Y then relevant at all? No. If we know, as in the above example, that the variation in the data only occurs over one dimension, Y can be neglected. This consideration can easily be transferred to much higher dimensions. All dimensions that do not contribute to the actual dynamics can theoretically be neglected—we just need to be able to identify them.

5.2.2   Eigenspace

How can we identify the most important dimensions and sort out those that are of less relevance to the dynamics covered in the data? For this, we use a well-known tool from linear algebra, the eigenvalue analysis. It forms the basis for our next steps.

We now briefly repeat the definition of eigenvalues and eigenvectors:

Vectors s_i, which, when transformed by a matrix $\mathbf{A}$ come to lie along their own direction again,

$$\mathbf{A}s_i = \lambda_i s_i. \tag{5.1}$$

are called **eigenvectors** of the matrix $\mathbf{A}$. The matrix thus does not act direction-changing, but like a linear factor λ_i. This factor is called **eigenvalue** of the matrix $\mathbf{A}$.

The real advantage of considering eigenvalues becomes clear when we now rewrite the matrix $\mathbf{A}$ with the help of knowledge of eigenvectors and eigenvalues:

Spectral decomposition. A square $N \times N$ matrix $\mathbf{A}$ can be decomposed into two matrices $\mathbf{S}$ and Λ such that

$$\mathbf{A} = \mathbf{S}\Lambda\,\mathbf{S}^{-1} \tag{5.2}$$

holds. Λ is a diagonal matrix that contains the eigenvalues λ_i of $\mathbf{A}$ on its diagonal. The matrix $\mathbf{S}$ contains in its i-th column the i-th eigenvector s_i.

Written out, Λ,

$$\Lambda = \begin{pmatrix} \lambda_0 & 0 & 0 & \dots & 0 \\ 0 & \lambda_1 & 0 & \dots & 0 \\ 0 & 0 & \lambda_2 & \dots & 0 \\ \vdots & \vdots & \vdots & \ddots & \vdots \\ 0 & 0 & 0 & \dots & \lambda_N \end{pmatrix} \tag{5.3}$$

where λ_i are the eigenvalues of the covariance matrix, while the eigenvectors s_i are columns in the matrix $\mathbf{S}$,

$$\mathbf{S} = \begin{pmatrix} s_{00} & s_{01} & s_{02} & \cdots & s_{0N} \\ s_{10} & s_{11} & s_{12} & \cdots & s_{1N} \\ s_{20} & s_{21} & s_{22} & \cdots & s_{2N} \\ \vdots & \vdots & \vdots & \ddots & \vdots \\ s_{N0} & s_{N1} & s_{N2} & \cdots & s_{NN} \end{pmatrix}. \tag{5.4}$$

> The space spanned by the eigenvectors s_i of a matrix A is called the **eigenspace** of the matrix A.

The spectral decomposition determines the eigenvalues and eigenvectors from the matrix $\mathbf{A}$ using the Jacobi method. This procedure is included in many libraries and programming languages and can therefore be easily applied to data today. We now return to our covariance matrix in more detail. The spectral decomposition also allows us to find suitable λ_i and s_i for this matrix.

5.2.3   Eigenspace of the Covariance Matrix

Let's start with a data matrix X. It contains our data vectors and thus the basis to establish the covariance matrix $\mathbf{C}(X)$, from which we can determine corresponding eigenvalues $\lambda_{C,i}$ and eigenvectors $s_{C,i}$. For our data matrix X and its row vectors x_i this means that we can transform them via $\mathbf{S}$:

> Every data vector x_i can be transformed into the eigenspace of the covariance matrix via
>
> $$\hat{x}_i = \mathcal{E}(x_i) = x_i \mathbf{S}_C \tag{5.5}$$
>
> where $\mathbf{S}_C$ is the matrix of eigenvectors of $\mathbf{C}$. We call this transformation **Karhunen-Loeve transform** or **Principal Component Analysis (PCA)**. We also refer to this process as **encoding** and indicate this by the designation $\mathcal{E}$.

We have another advantage in relation to $\mathbf{S}$, which comes directly from the properties of the covariance matrix itself—$\mathbf{C}$ is symmetric due to its definition. For symmetric matrices, the spectral decomposition is further simplified, as the inverse of $\mathbf{S}, \mathbf{S}^{-1}$, can be determined directly from the transposition of the matrix,

$$\mathbf{S}^{-1} = \mathbf{S}^T, \tag{5.6}$$

which leads to the spectral decomposition of the covariance matrix

$$\mathbf{C} = \mathbf{S} \Lambda \mathbf{S}^T \tag{5.7}$$

and makes the reverse transformation of the data similarly simple:

> A vector $\hat{x}_i$ from the eigenspace of $\mathbf{C}$ can be transformed back into the data space via
>
> $$x_i = \mathcal{D}(\hat{x}_i) = \mathbf{S}_C^T \hat{x}_i \tag{5.8}$$
>
> In the case of $\mathcal{D}$ we also speak of **Decoding.**

With the formalisms (5.5) and (5.8) we can jump into the eigenspace and back again. Boardman et al. demonstrated in several works [1, 2] how well PCA can be applied to pattern recognition problems in spectra. They use the above transformation to search for elementary properties of hazardous substances in the eigenspace.

5.2.4   Eigenspace Reduction

Our initial example, however, dealt with the reduction of dimensions. So far, we have only transformed into an equally dimensioned, albeit different space with the transformation (5.5). Therefore, a step is still missing to reduce the dimensions.

We now again consider the matrix of eigenvectors of $\mathbf{C}$. It consists of N eigenvectors, of which we assume that only $L < N$ really contribute to the dynamics. We therefore eliminate all further eigenvectors from $\mathbf{S}_C$ and thus arrive at a new, more compact matrix $\tilde{\mathbf{S}}$,

$$\tilde{\mathbf{S}} = \begin{pmatrix} s_{00} & s_{01} & s_{02} & \cdots & s_{0L} & \diagdown & s_{0N} \\ s_{10} & s_{11} & s_{12} & \cdots & s_{1L} & \diagdown & s_{1N} \\ s_{20} & s_{21} & s_{22} & \cdots & s_{2L} & \diagdown & s_{2N} \\ \vdots & \vdots & \vdots & \ddots & \cdots & \diagdown & \vdots \\ s_{N0} & s_{N1} & s_{N2} & \cdots & s_{NL} & \diagdown & s_{NN} \end{pmatrix}. \tag{5.9}$$

This process reduces the dimension. We also call the leading eigenvectors the principal components of our data matrix $\mathbf{X}$. Often in practice, only a few of these principal components are necessary to fully capture the data.

Once the matrix is reduced, we can also switch back to the actual data space, for this we only need to apply (5.10):

> PCA allows for a **linear compression** of the relevant information by eliminating irrelevant eigenvectors. The dimension is effectively reduced. The low-dimensional data vectors are given by,
>
> $$\tilde{x}_i = \tilde{\mathbf{S}}_C^T \hat{x}_i,. \tag{5.10}$$

Where we use the symbol $\tilde{x}_i$ to show that it is now explicitly not the data vector x_i itself, but its low-dimensional reconstruction obtained from $\tilde{\mathbf{S}}$.

5.2.5   *Implementation via Scikit-Learn*

Here we show a particularly simple and fast implementation using the Scikit-Learn package. For this, we first need to import the decomposition PCA from sklearn. In Listing 5.1, our familiar example with the motor current curves is then loaded and colored using known labels. This process is only necessary for our better view of the data. We can then better distinguish the transformations later. In Fig. 5.2(a) these colored raw data can be seen. About 1000 curves have been plotted, the blur is caused by a low alpha value per curve.

Listing 5.1 Preparing example data for the PCA

```
import matplotlib.pyplot as plt
import pickle
import numpy as np
from sklearn.decomposition import PCA

data = pickle.load(open('EX03Engine.pickle','rb'))
X = data['X']
mu = np.mean(X, axis=1)

mycolor=[]
for i in range(0,500):
    if data['Label'][i] == 1:
        mycolor.append('r')
    elif data['Label'][i] == 2:
        mycolor.append('b')
    elif data['Label'][i] == 3:
        mycolor.append('y')
    else:
        mycolor.append('k')
```

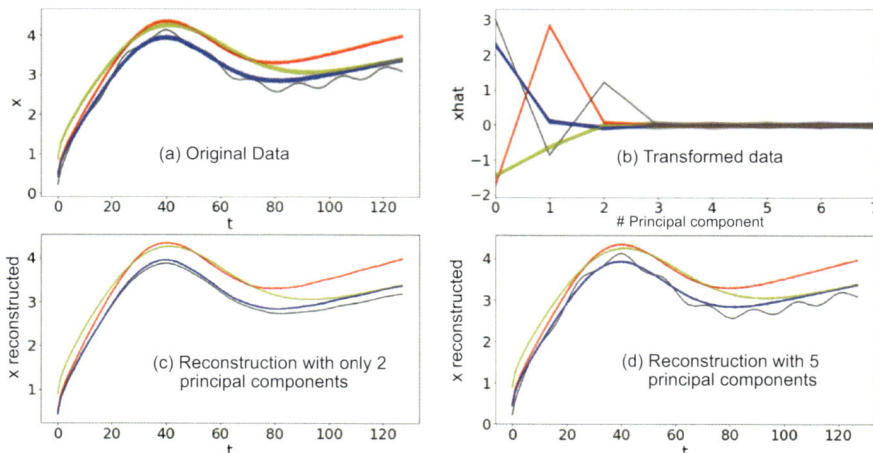

Fig. 5.2 Principal component analysis using our example with motor currents. **a** Representation of the original input data of the PCA, **b** transformed data in the eigen space, 8 main components are shown, **c** poor reconstruction considering a transformation with only 2 main components and **d** good reconstruction with 5 main components

In Listing 5.2, the PCA is applied to the data in X and the new matrix Xhat is calculated. Please note line 8. Here, the mean value of each data series is stored in a dedicated variable mu. For the evaluation of the PCA, the mean value is subtracted by the method. We therefore need this variable to get back to the original space later.

Listing 5.2 Implementation of the PCA

```
# Transformation to principal components
pca = PCA()
pca.fit(X)
Xhat = pca.transform(X)
```

The result of Listing 5.2 is shown in Fig. 5.2(b). It is the abstract transformation of data vectors into the eigenspace. Only the first 8 components are shown here, as the dynamics can apparently be fully captured between component 0 and 4. For higher components > 5 we see only noise in Fig. 5.2(b).

Listing 5.3 Backtransformation

```
nComp = 5
Xtilde = np.dot(Xhat[:,:nComp], pca.components_[:nComp, :])
Xtilde = Xtilde+mu
```

In Listing 5.3, the actual dimension reduction is now carried out. However, we use the plane of $\hat{\mathbf{X}}$ and eliminate all dimensions larger than nComp. The reconstruction quality that we achieve in this way is shown in Fig. 5.2(c) for only 2 components and in Fig. 5.2(d) for 5 components. While 2 components are too few to fully reconstruct the data—among other things, we lose the oscillation information of the anomaly! -, a reconstruction with 5 components can very well reproduce the basic courses.

Another observation is the effective denoising of the original data. After reconstruction, the curves appear smoother and sharper. This effect comes from cutting the noise components in the eigenspace.

5.2.6   Discussion of the PCA

The PCA can be applied to a wide variety of problems and as we have seen, it can effectively reduce dimensions. PCA transformations are excellent for further processing in supervised learning methods such as neural networks and decision trees.

The PCA is also well suited for preprocessing. There are now two types of data with which you can proceed:

1. You can truncate the encoded data $\hat{x}$ of their irrelevant parts and connect further analyses in the eigenspace.
2. You can return to the original data space via the decoding in (5.8) and apply subsequent procedures from here.

In both cases, the dimension is reduced.

Application for anomaly detection
Shyu et al. show in [11] how the PCA can be used for anomaly detection. This work is representative of a multitude of such use cases. If an irregularity is found in the eigen space that deviates from normal behavior, then there is an anomaly. This is often easier to detect in the eigenspace than in the original data space.

Here too, caution is required and the results should be carefully scrutinized. If we revisit the example with the marbles, we said that the variation along Y does not matter and only the dimension X reflects the entire variance of the data. In a consistent reduction, the influence in Y would eventually be neglected. But this dimension could be the anomaly.

Therefore, it only makes sense to develop an anomaly detection with a PCA if you already have prior knowledge about the anomalies you want to find. They must be contained in the set of data that we use to build the covariance matrix.

Remarks on application
The choice of the evaluator space is often problem-dependent, but we would like to provide some guidance for application with technical process data:

- The PCA denoises your data extremely effectively. In combination with the moving average filter, information can therefore be lost.

- As can be seen in the example and in Fig. 5.2, the oscillation of the original data is lost if we reduce the dimension too much or if we remain in the eigenspace.
- If you want to analyze oscillations in data, use the PCA before an FFT and make sure that the data actually contain the relevant frequencies. In this case, denoising via the PCA helps to significantly improve and sharpen the identification of frequencies in the FFT spectrum.

Disadvantages
However, the PCA also has a weakness: It is a linear method. The projection into the eigen space is linear and therefore not all data courses can be well transformed with the PCA. You should try to apply the PCA to your individual problems through targeted testing and decide on a case-by-case basis how well the situation is reconstructed. For this reason, we have explained the reconstruction method in more detail above. It can help you determine how well the method works.

In this chapter, we will get to know a relative of the PCA, the autoencoder, which can be interpreted as a nonlinear principal component transformation. This method is more complex, but applicable in almost all situations.

5.3   K-Means Clustering Method

5.3.1   Finding Cluster Centers

The K-Means—K-Means—Algorithm is one of the most widely used cluster algorithms. Its operation is simple, understandable, and it shows robust results even in demanding situations. Its biggest disadvantage is the prior knowledge, or estimation, of the number of suspected cluster centers.

We will now discuss the individual steps of the K-Means algorithm. For this, we assume an example in two dimensions. In this manageable size, we can explain the procedure more easily. In general, however, the dimensions are not limited. In Fig. 5.3 we have also sketched the individual steps.

- **(A) Random positioning.** k random position candidates are chosen. These centers are given in our simple example by vectors m_R, m_B, and m_G, where R stands for Red, B for Blue, and G for Green. In this case, $k = 3$. In Fig. 5.3-1, the positions are represented by correspondingly colored centers of circles.
- **(B) Summarizing and Assigning.** Starting from N existing data points, each data point x_i is now successively assigned to one of these centers. To do this, we again minimize a cost function

$$J = \sum_{j=0}^{k-1} \sum_{i}^{N} ||x_i - m_j||^2, \tag{5.11}$$

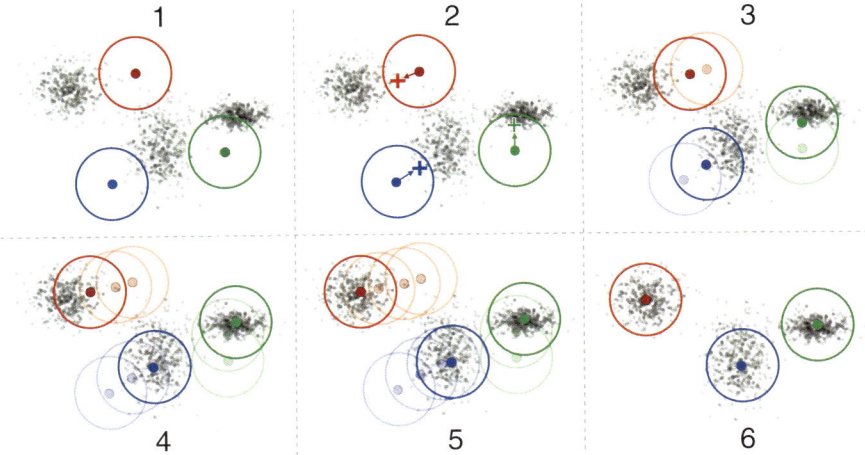

Fig. 5.3 Schematic sequence of the K-Means algorithm

by checking for each data point every possible assignment to a center m_j and selecting the assignment that increases J the least. This cost function is also referred to as the variance in the cluster.

- **(C) Finding centroids.** Once all data points are assigned, there are k sets of data points. Now we calculate the centroids of these data sets. In Fig. 5.3-2, these centroids are marked as colored crosses. As a final step, we define these centroids as new candidates for cluster centers and repeat the procedure from (B).

Fig. 5.3 shows over several iterations how the candidates converge into the cluster centers using the algorithm described above.

5.3.2   Implementation with Scikit-Learn

The implementation of the K-Means method is as simple as the above explanation. But before we can examine the method in Python code, we need to think of a test example that helps us understand the clustering. We use the point clouds already shown in Fig. 5.3. We generate these point clouds in Listing 5.4 using 2D Gaussian functions. Lines 6, 7, and 8 lead to clustered data points (x_i, y_i).

Listing 5.4 Generating synthetic data for the K-Means method

```
def gauss2d(mu_x, sigma_x, mu_y, sigma_y, numberOfPoints=400):
    x1 = mu_x + sigma_x * np.random.randn(numberOfPoints)
    y1 = mu_y + sigma_y * np.random.randn(numberOfPoints)
    return x1, y1

x1, y1 = gauss2d(5, 1, -4, 3)
x2, y2 = gauss2d(-1, 1, 5, 2)
x3, y3 = gauss2d(10, 1, 1, 1)
```

Often, you do not have to go this way on your own. Scikit-Learn also provides various tools for generating synthetic data. We prefer our own implementation here, as this allows you to always make your own extensions or adjustments to the routine. We convert the data points of the three point clouds into a training set `Xtrain`, and insert this into the Scikit-Learn K-Means algorithm. Listing 5.5 shows the application of the algorithm itself.

Listing 5.5 Applying the K-Means algorithm

```
from sklearn.cluster import KMeans

Xtrain = []
for i in range(0, 300):
    Xtrain.append([x1[i],y1[i]])
    Xtrain.append([x2[i],y2[i]])
    Xtrain.append([x3[i],y3[i]])

kmeans = KMeans(n_clusters=3, random_state=0)
kmeans.fit(np.array(Xtrain))

print(kmeans.cluster_centers_)
```

As the last point of this code, the cluster centers are output in line 12. Now we check how well the trained cluster assignment maps our data points. We formulate this test in Listing 5.6. First, the test set `Xtest` is generated from the remaining data points. We apply the previously trained method to each data point using `kmeans.predict()`.

Listing 5.6 Plotting the K-Mean results

```
Xtest = []
for i in range(301, 400):
    Xtest.append([x1[i],y1[i]])
    Xtest.append([x2[i],y2[i]])
    Xtest.append([x3[i],y3[i]])

for eachElement in Xtest:
    print(eachElement)
    result = kmeans.predict(np.array(eachElement).reshape
        (1,-1))
    if result == 0:
        color = 'r'
    elif result == 1:
        color = 'b'
    else:
        color = 'g'

    plt.scatter(eachElement[0], eachElement[1], s=20, color=
        color)
```

The visualization now shows an automatic assignment into the clusters with red, blue, and green. The choice of colors and positions of the clusters correspond to the example in Fig. 5.3.

5.3.3   Batch K-Means

The algorithm shown here has some disadvantages in terms of its strategy on very large data sets. In the above step (B), the method must perform the minimization step for each data point. This means that the algorithm internally uses various loops to calculate the sums of (5.11). While the first sum is manageable, as it only involves a previously known number of cluster centers, the second sum is critical as it increases with the number of points.

To avoid this weakness, there are advanced approaches that search for local optima [4] or search on smaller subsets [10]. For this, random subsets of X are used in each iteration of the algorithm. Although they represent the original data set, they contain a limited, predefined number of points. Since only this subset is evaluated, the resulting algorithm is faster.

In Scikit-Learn, this approach exists under the name Mini Batch K-Means. The application of this algorithm variant to our example is analogous to Listing 5.5 and is listed in Listing 5.7.

Listing 5.7 Applying Mini-Batch-K-Means

```
from sklearn.cluster import MiniBatchKMeans, KMeans

miniBatchKmeans = MiniBatchKMeans(n_clusters = 3, batch_size =
    20, n_init = 10)
miniBatchKmeans.fit(Xtrain)
```

5.4   The t-Distributed Stochastic Neighbour Embedding Algorithm (t-SNE)

5.4.1   Idea behind t-SNE

The t-SNE algorithm analyzes the abstract distances of data points to get a measure of how likely two points are to be adjacent. Abstract, because the term distance here is also to be understood in the context of similarity or proximity. The method is based on the work of G. Hinton and S. Roweis, who in 2003 [3] dealt intensively with the neighborhood of data points and introduced the Stochastic Neighbour Embedding. L. v. Maaten and G. Hinton finally demonstrated the use of the t-Student distribution in [6], which led to the algorithm in its current form.

If we consider the data points $x_1, x_2, \ldots, x_N$, this proximity is defined by a conditional probability $p(i|j)$, where i and j denote the indices of our points. We also use vectors $x_i \in \mathbb{R}^D$, as each data point can be represented by a sufficiently large set D of dimensions.

In other words, $p(i|j)$ expresses the probability that a point $\boldsymbol{x}_i$ is near the point $\boldsymbol{x}_j$. For this, we need to make an assumption, namely which distribution we want to use here, and t-SNE uses the Gaussian distribution. For the conditional probability, we define

$$p(i|j) = \frac{1}{Q} \exp\left[-\frac{||\boldsymbol{x}_i - \boldsymbol{x}_j||^2}{2\sigma^2}\right] \tag{5.12}$$

where the auxiliary size Q applies

$$Q = \sum_{k \neq i} \exp\left[-\frac{||\boldsymbol{x}_i - \boldsymbol{x}_k||^2}{2\sigma^2}\right] \tag{5.13}$$

With $||\boldsymbol{x}_i - \boldsymbol{x}_j||$ we determine the distance between the data vector $\boldsymbol{x}_i$ and $\boldsymbol{x}_j$. Due to the fact that $\boldsymbol{x}_i$ are our input data, we can calculate (5.12) for between all pair combinations.

In the next step, the t-SNE method performs a dimension reduction: it maps the data points into a low-dimensional space $\mathbb{R}^d$, where $d \ll D$ is and is often chosen with $d = 2$. The choice of $d = 2$ allows us to visualize the data in the target space on a surface.

For the points $\boldsymbol{y}_i \in \mathbb{R}^d$ in the target space, t-SNE finally also requires a neighborhood distribution $\tilde{p}(i|j)$, defined as,

$$\tilde{p}(i|j) = \frac{1}{\tilde{Q}}\left[\frac{1}{1 + ||\boldsymbol{y}_i - \boldsymbol{y}_j||^2}\right] \tag{5.14}$$

with the auxiliary size $\tilde{Q}$

$$\tilde{Q} = \sum_l \sum_{l \neq m} \frac{1}{1 + ||\boldsymbol{y}_l - \boldsymbol{y}_m||^2}. \tag{5.15}$$

As a final step, the Kullback-Leibler divergence

$$\mathrm{KL}(p, \tilde{p}) = \sum p(i|j) \log \frac{p(i|j)}{\tilde{p}(i|j)} \tag{5.16}$$

between both distributions is minimized. What happens in a descriptive way? The neighborhood relationships from the space of the input data are preserved, as the divergence between the distributions becomes small. Through the special form of $\tilde{p}$, the data is additionally distributed so that neighboring points move close together and clusters take up as much distance as possible from each other.

5.4.2   Application of the t-SNE Implementation in Scikit-Learn

Scikit-Learn includes an implementation of t-SNE. It can be applied very quickly to practical problems. We demonstrate this here using our time series example for motor currents. The complete listing is shown in 5.8. There, the example dataset is loaded and colored again. This use of labels is intended only for illustration. The labels themselves are not used in the procedure.

Listing 5.8 Applying t-SNE to the engine data example

```
import matplotlib.pyplot as plt
import numpy as np
import pickle
from sklearn.manifold import TSNE

data = pickle.load(open('EX03Engine.pickle', 'rb'))
x = data['X'][0:500]

mycolor=[]
for i in range(0, len(x)):
    if data['Label'][i] == 1:
        mycolor.append('r')
    elif data['Label'][i] == 2:
        mycolor.append('b')
    elif data['Label'][i] == 3:
        mycolor.append('y')
    else:
        mycolor.append('k')

myTsne = TSNE()
y = myTsne.fit_transform(x)

for i in range(0,500):
    plt.scatter(y[i,0],y[i,1], color = mycolor[i], alpha=0.2)

plt.xlabel('$y_0$', fontsize=22)
plt.ylabel('$y_1$', fontsize=22)
plt.tick_params('both', labelsize=20)
```

In Fig. 5.4 we show in (a) again the time series, colored according to their label, and in (b) the t-SNE result of the above code. In addition, it was marked how the time series transform into the different clusters. Please pay close attention to the black point in the top right. This point summarizes the occurrences of the anomaly that occurs multiple times in the data. In our discussion of the FFT in Sect. 3.6.3 the same example was considered and the anomaly was identified by its frequency. The t-SNE result also tells us that this special case must be very close to the blue curve.

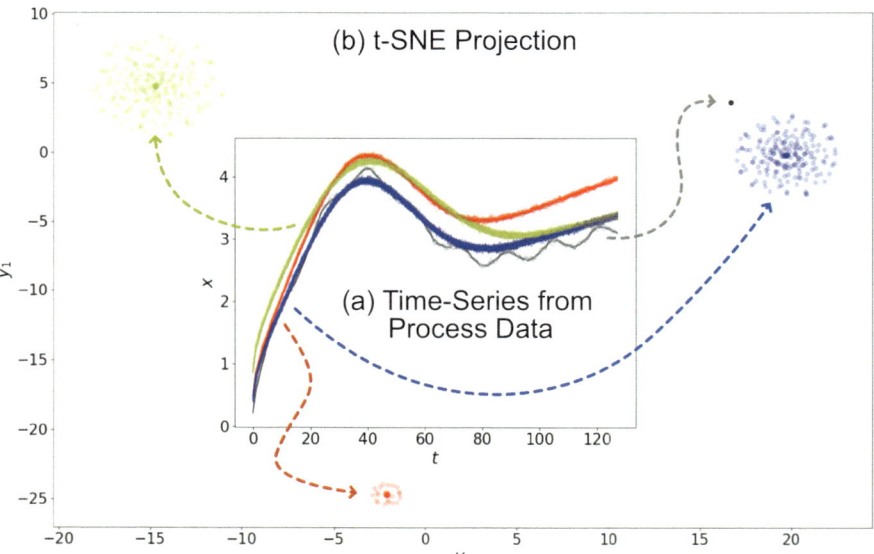

Fig. 5.4 Application of t-SNE to the motor current example. Dashed lines symbolize the assignment of the time series in (a) to the cluster representation in the t-SNE projection plane (b)

5.4.3   Advantages and Disadvantages of t-SNE

So if we were working on unknown data, we would gain information and a good overview of our data through the application of t-SNE. However, there are also disadvantages to this procedure:

- **Execution speed.** The t-SNE procedure works slowly as it has to iterate through all pairs of points. Therefore, it is only partially usable in the online case.
- **Low reproducibility.** Due to the inherent starting randomness, the distributions of the cluster positions are different with each execution.

Many of these deficits are no longer present in advanced procedures. An example is the Uniform Manifold Approximation and Projection (UMAP), which was introduced by McInnes in [7].

5.5   Autoencoder

5.5.1   Topology of an Autoencoder

An autoencoder (AE) is a neural network with a special topology: The input variables are simultaneously the training targets. So we feed data from the left into the network and expect, after successful training, that this input is reproduced as well as possible at the output. The AE is thus a 1-mapping.

> A neural network, for which applies
>
> $$z = \mathcal{A}(x) \text{ mit } z \xrightarrow[\text{Train}]{} x \tag{5.17}$$
>
> is called **Autoencoder.**

If you divide the network at a middle layer, you can identify two half networks, the encoder and the decoder part.

> If we write (5.17) as
>
> $$z = \mathcal{D}\big[\mathcal{E}(x)\big] \tag{5.18}$$
>
> we identify $y = \mathcal{E}(x)$ as the **Encoder** and $z = \mathcal{D}(y)$ as the **Decoder** of the autoencoder.

The special application of such a network is based on its special topology, which is shown in Fig. 5.5. The input and output layers are identically dimensioned as explained above. The middle, hidden layers of the network, however, were chosen so that their number of neurons decreases to a certain level (in the example 3 neurons). After that, the number of neurons in the layers increases again. If such a network is now trained to map x onto x, the information contained in x must be compressed by the network to the small number of middle neurons. This is exactly what makes the AE.

The dimension of the input is effectively reduced to the dimension of the middle layer. Thus, we have determined a property of the AE:

> Encoders of autoencoders reduce the dimension of the input variables through **nonlinear compression** in the hidden layers.

While a PCA is a linear transformation, the autoencoder allows for **nonlinear transformation** and can better adapt to complex boundary conditions.

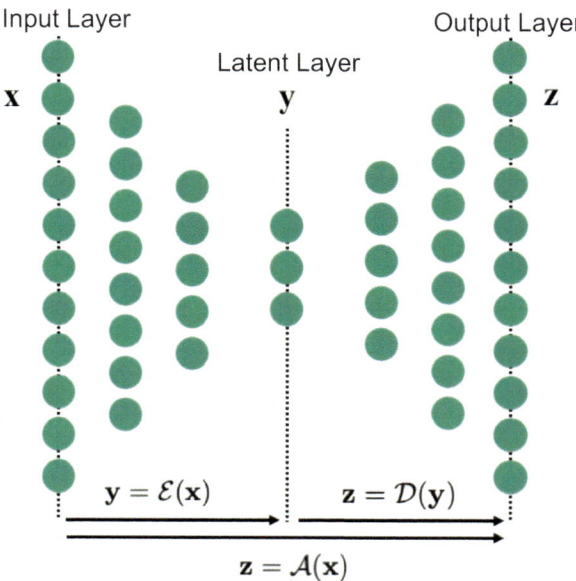

Fig. 5.5 Illustration of the topology of an autoencoder

5.5.2   Latent Space

The middle layer of the Autoencoder thus contains a nonlinearly compressed representation of our input data. It plays a special role, similar to the eigenspace of the covariance matrix.

> If there exists an autoencoder $\mathcal{A}$, for which $x \approx \mathcal{A}(x)$ applies, and if $\mathcal{E} : \mathbb{R}^N \leftarrow \mathbb{L}$ is the trained encoder part of $\mathcal{A}$, we call $\xi \in \mathbb{L}$ the **encoded representation** of x and $\mathbb{L}$ the **latent space** of the autoencoder.

It is permissible to speak of an eigenspace here as well. Strictly speaking, $x = \mathcal{A}(x)$ is an eigenvalue equation, where after training $\mathcal{A} \approx \mathbf{1}$ is and each x_i of the input data represents an eigenvector.

5.5.3 Anomaly Detection

A of the most common applications for autoencoders is anomaly detection. Various works, [5, 9] or [8], discuss application examples. We would therefore like to discuss the basic strategies with which an autoencoder identifies an anomaly:

- **Quality of Transformation.** The autoencoder is trained on a data set X, later in the code `Xtrain`. It therefore reconstructs all types of curves x_i very well in terms of

$$x_i \approx \mathcal{A}(x_i). \tag{5.19}$$

If a reconstruction is poor, $x_i \neq \mathcal{A}(x_i)$, it is a type of input data that was not trained. A simple comparison, for example with a least-square approach,

$$J = |\mathcal{A}(x_i) - x_i|^2 \tag{5.20}$$

provides information about the quality of reconstruction and can even quantify it over J.
- **Anomaly detection in the latent space.** Alternatively, one looks at the data distribution in the latent space $\mathbb{L}$. Here, the normal cases arrange themselves in clusters. Anomalies stand out because they move away from these groups of normal cases.

Please pay close attention to these statements. The autoencoder not only detects an anomaly. It is able to tell you whether it has seen the data before or not. It therefore identifies whether a behavior of the input variables is fundamentally unknown or known. The anomaly itself does not have to have been part of the training data set. Training on a series of normal cases is sufficient.

5.5.4 Implementation in Keras

We use our neural networks from the previous chapter to implement an AE in Keras. For this, we want to use the topology from Fig. 5.5. However, before we get to the actual network, we need to think of a use case. Here, simple pattern recognition is suitable. We define Gaussian pulses with different widths and positions. We want to distinguish these with an AE.

Labels are not needed, as this is an unsupervised method. The AE must find the best representation in its latent layer itself. Nevertheless, we also generate labels. They are not used for training, but are needed later for testing.

The implementation of our synthetic data generation is shown in Listing 5.9.

Listing 5.9 Definition of synthetic training and test data for the Autoencoder

```python
def gaussianDistribution(t, mu, sigma):
    return np.exp( -(t-mu)**2 / 2 / sigma**2 )

def generatePulses(numberOfEvents=10):
    X = []
    Y = []
    t = np.array(range(0,100),dtype=float)
    r = np.random.random(numberOfEvents)
    for i in range(0,numberOfEvents):
        if r[i] > 0.5:
            thisSigma=8
            thisMu = 70
            thisLabel=0
        else:
            thisSigma=2
            thisMu = 10
            thisLabel=1
        f = gaussianDistribution(t, mu=thisMu, sigma=thisSigma
            )

        X.append(f)
        Y.append(thisLabel)
    return X, Y

Xtrain, Ytrain = generatePulses(5000)
Xtest, Ytest = generatePulses(50)
```

A random process alternates between two Gaussian shapes. For training, a set Xtrain is generated. Ytrain is a placeholder. Xtest and Ytest are test data, where we primarily use Ytest for visualization and checking our results.

The following Listing 5.10 shows the actual programming of the AE as a Keras model. We have chosen a topology of the form 100,20,10,3,10,20,100. So that you have direct access to the encoder and decoder part of the autoencoder with this example code, the model evaluation has already been set up accordingly in lines 28 and 29.

Listing 5.10 Implementation of an Autoencoder in Keras

```
import tensorflow as tf
import keras
from tensorflow.keras import layers, losses
from tensorflow.keras.models import Model

class Autoencoder(Model):

    def __init__(self, latent_dim):
        super(Autoencoder, self).__init__()
        self.latent_dim = latent_dim

        self.encoder = tf.keras.Sequential([
          layers.Input(100), ## 128 for AE, 100 for Gauss
          layers.Dense(20, activation='leaky_relu'),
          layers.Dense(10, activation='leaky_relu'),
          layers.Dense(self.latent_dim, activation='sigmoid'),
        ])
        self.decoder = tf.keras.Sequential([
          layers.Dense(10, activation='leaky_relu'),
          layers.Dense(20, activation='leaky_relu'),
          layers.Dense(100, activation='sigmoid') ## 128 for AE
              , 100 for Gauss
        ])

        self.compile(optimizer='adam',loss=losses.
            MeanSquaredError())
        self.optimizer.learning_rate = 0.005
    def call(self, x):
        encoded = self.encoder(x)
        decoded = self.decoder(encoded)
        return decoded

autoencoder = Autoencoder(3)
history = autoencoder.fit(np.array(Xtrain), np.array(Xtrain),
    epochs=5, batch_size=25, shuffle=True)
```

We train the AE with Xtrain at the input and output. Thus, the input is mapped to the output. The training for this particular case should not take much time. Ultimately, we are again faced with the problem of checking our model. Here we choose a visualization that shows the input, the latent layer, and finally the reconstruction at the output.

Fig. 5.6 shows the input of 10 tests in the top row, the latent layer in the middle (red bars), and the reconstructed result at the output. The latent layer shows the encoding result, which in this case means that there are two compressed states: a) all three neurons excited leads to the wide Gaussian at the upper position and b) only the first neuron excited, while the following two neurons are zero, which is associated with the narrow, front Gaussian.

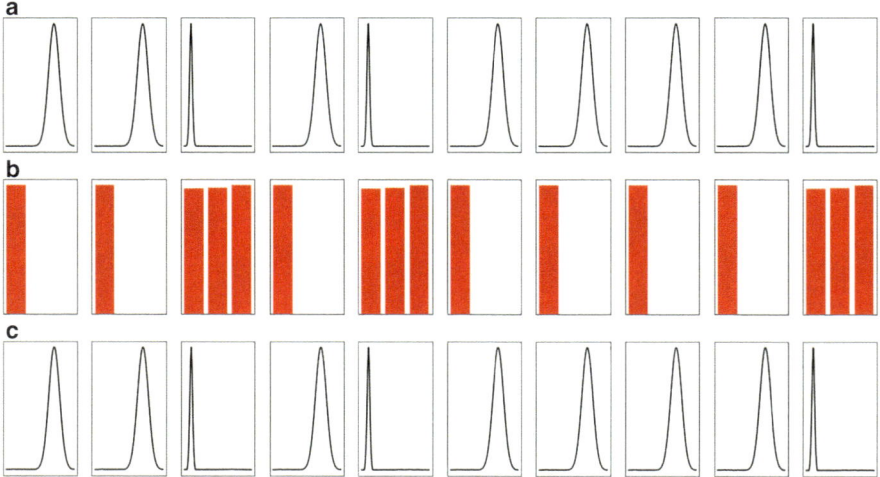

Fig. 5.6 Result of the AE from Listing 5.10, represented with the visualization code from Listing 5.11

Listing 5.11 Validation of the AE results

```
encodedGauss = autoencoder.encoder(np.array(Xtest)).numpy()
decodedGauss = autoencoder.decoder(encodedGauss).numpy()
n = 10
plt.figure(figsize=(11, 7),dpi=100)
for i in range(n):
    ax = plt.subplot(3, n, i + 1)
    plt.plot(Xtest[i], 'k', linewidth=2.0)
    plt.gray()
    ax.get_xaxis().set_visible(False)
    ax.get_yaxis().set_visible(False)

    ax = plt.subplot(3, n, i + 1 + n)
    plt.bar(range(0,len(encodedGauss[i])), encodedGauss[i],
        color='r',linewidth=2)
    plt.gray()
    ax.get_xaxis().set_visible(False)
    ax.get_yaxis().set_visible(False)

    ax = plt.subplot(3, n, i + 2*n+1)
    plt.plot(decodedGauss[i], 'k', linewidth=2.0)
    plt.gray()
    ax.get_xaxis().set_visible(False)
    ax.get_yaxis().set_visible(False)
plt.show()
```

5.5.5 Classification in the Latent Space

Since the latent space of the AE has a special significance, it makes sense to specifically output these neurons. In Listing 5.12 we use this approach to represent the distribution of neuron activities.

Listing 5.12 Visualization of the latent neurons

```
encodedGauss = autoencoder.encoder(np.array(Xtest)).numpy()
decodedGauss = autoencoder.decoder(encodedGauss).numpy()

n = 10
plt.figure(figsize=(11, 7),dpi=100)
for i in range(n):
    if Ytest[i]==0:
        myColor = 'r'
    else:
        myColor = 'k'
    plt.scatter(range(0,len(encodedGauss[i])), encodedGauss[i
        ], s=120, color=myColor)

plt.grid(True)
plt.xlabel('#_Neuron_der_latenten_Schicht', fontsize=22)
plt.ylabel('Anregung_des_Neurons', fontsize=22)
plt.tick_params('both', labelsize=20.0)
plt.xticks([0,1,2])
plt.show()
```

Please note that we only use `Ytest` here to generate the color. It helps us to examine the different cases. Fig. 5.7 shows the visualization of the latent layer.

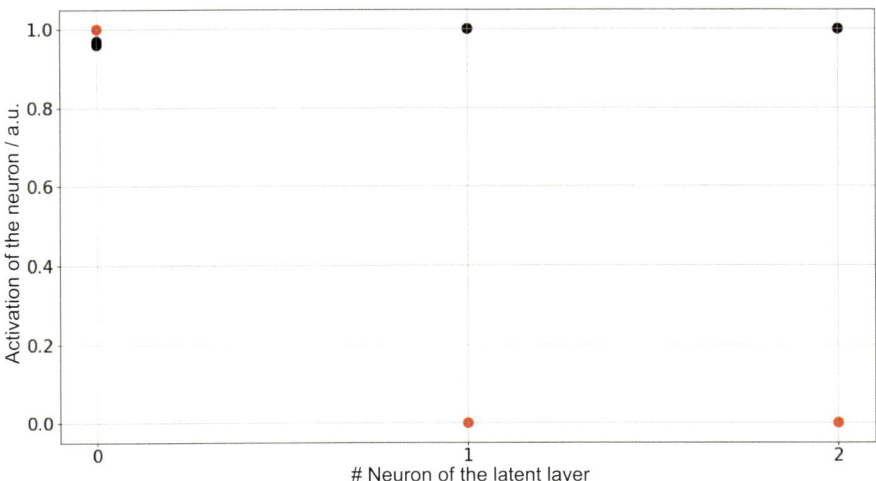

Fig. 5.7 Visualization of the latent space

Using our example, we can see that a simple algorithm can now check which Gaussian shape is present. Thus, the simple conditions

$$a_1 < 0.5 \quad \text{oder} \quad a_3 < 0.5, \tag{5.21}$$

could already bring about the group's decision. At this point, the following steps are suggested:

- Formulation of a simple decision-making procedure like (5.21),
- Use of a downstream neural classification procedure,
- Use of a decision tree.

Through the result of the AE, we have labels for the supervised learning methods. Therefore, a subsequent procedure can be used to move from the unsupervised learned latent space to a classification.

5.5.6   Uncertainty in the AE

We continue to consider our test case with Gaussian functions. This case is of course much simpler than the situations we expect in reality. However, it is excellent for observing how an AE behaves under disturbances and uncertainties.

For the moment, we leave the training exactly as we achieved it above in Listing 5.10. However, we modify the generation of our test data and add the following random process in line 18 of Listing 5.9:

Listing 5.13 Changing the test data
```
18  f = gaussianDistribution(t, mu=thisMu, sigma=thisSigma +0.1*np
        .random.randn(100)
```

Fig. 5.8 shows the results of the noisy case. The reconstruction is still good. The latent neurons continue to classify the results correctly. A noise of 10 % thus does not cause any disturbance in the assignment. Since we are still using the AE, which was trained on the unnoisy Gaussian functions, the AE tries to reproduce exactly these.

5.5.7   Detection of an Unknown Anomaly

Preparation of the test data
   We now want to investigate how to detect a fundamentally unknown anomaly. For this purpose, we modify the code for generating our training and test data once again. First, we increase the noise to 25 %, in the same line as before:

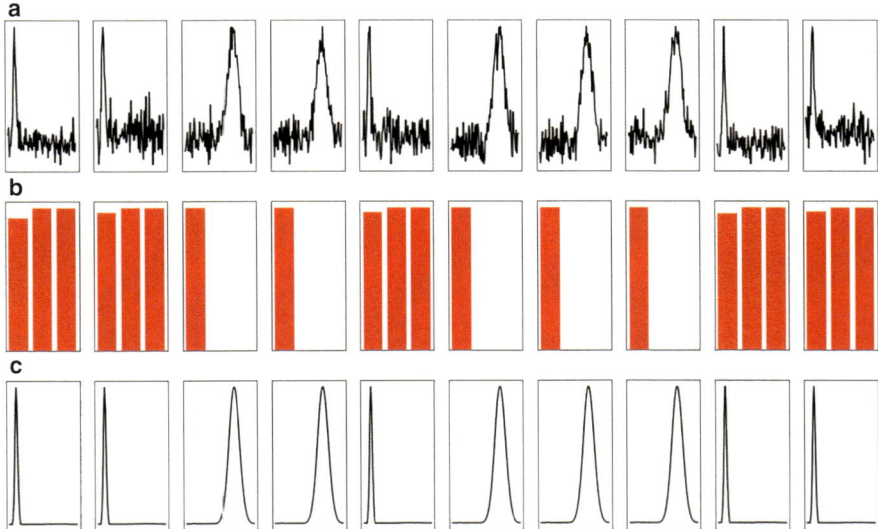

Fig. 5.8 Test of the AE with noisy input data

Listing 5.14 Changing the test data by adding noise

```
18  f = gaussianDistribution(t, mu=thisMu, sigma=thisSigma +0.25*
        np.random.randn(100)
```

Afterwards, we finally add the following under line 25 in the code 5.9:

Listing 5.15 Changing the test data by adding a gaussian peak

```
18  Xtest[5] = gaussianDistribution(np.array(range(0,100),dtype=
        float), mu=50, sigma=2)+0.25*np.random.randn(100)
```

Thus, we add exactly one exotic case to `Xtest`: a Gaussian form at a completely different position than all other cases. We do not add this case to the training set! This is important, the AE should know nothing about this case during training—otherwise it would be a known anomaly.

Detection in the latent space

Previously, we in Sect. 5.5.3 had shown two ways how we can detect an anomaly. One way examines the distributions of known data in the latent space. We will start with this approach. We train the AE with the code from Listing 5.10, using the noisy data and examine the result with Listing 5.11. Fig. 5.9 shows the comparison of input, latent layer and output. The compressed result in the latent layer of the sixth test case is different from the other cases. Here we see the effect of our artificial anomaly. We also recognize how the AE behaves in the reconstruction. It tries to reconstruct a case it knows.

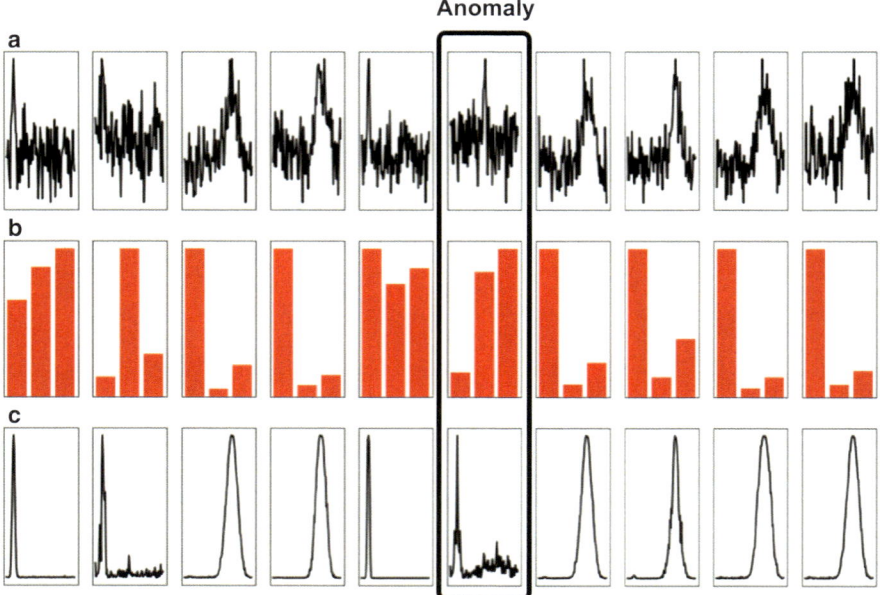

Fig. 5.9 AE input, latent layer and output, for 10 tests on the noisy data with a dedicated anomaly added at the position $i = 5$ (the sixth entry)

We overlay this single case in the plot of all test cases. For this purpose, an additional, single scatter plot point of our test case is added. Fig. 5.10 shows the result and allows us to analyze the position of the anomaly in the activation layers. In a real dataset, where no labels are known, the decision for the presence of an anomaly must be made solely on the basis of the activations and the reconstruction.

We can go one step further and draw the distributions of neuron activation, which can also be seen in Fig. 5.10, as histograms. Fig. 5.11 shows the histograms for each of the three latent neurons. With the help of the labels, we have colored the areas to better assign them to the two classes of Gaussian functions. A single, blue bar symbolizes the anomaly. The latter can be clearly distinguished from the normal cases.

Conditional probability in the latent space

The conditional probability helps us to further optimize the prediction of an anomaly. It is not just the probability whether A_2 e.g. belongs to an anomaly; rather, we must ask whether A_2 belongs to an anomaly, under the condition that A_1 applies. The diagrams in Fig. 5.11 only show us the distribution of the activation and are thus proportional to the probability of finding an activation at a certain numerical value.

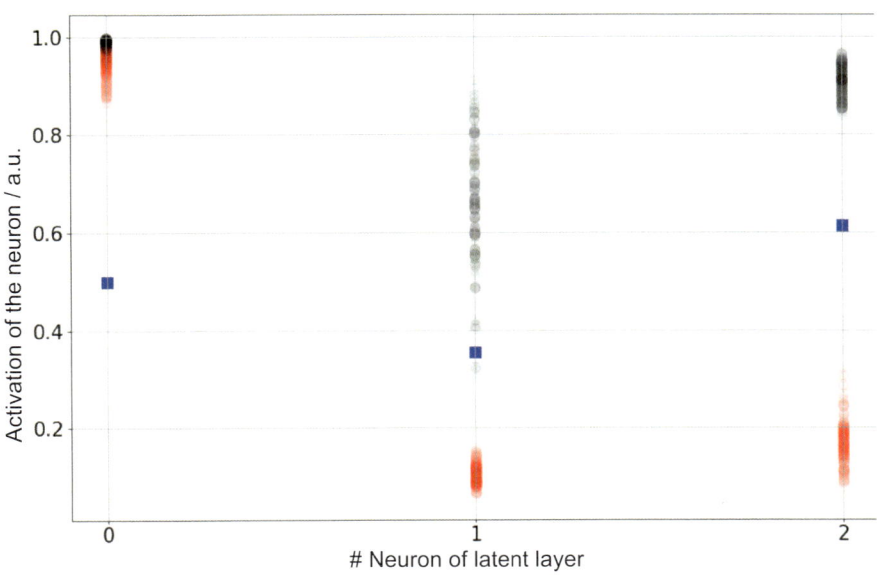

Fig. 5.10 AE result on the noisy data with a dedicated anomaly added. The anomaly shown in Fig. 5.9 is entered here as a blue square

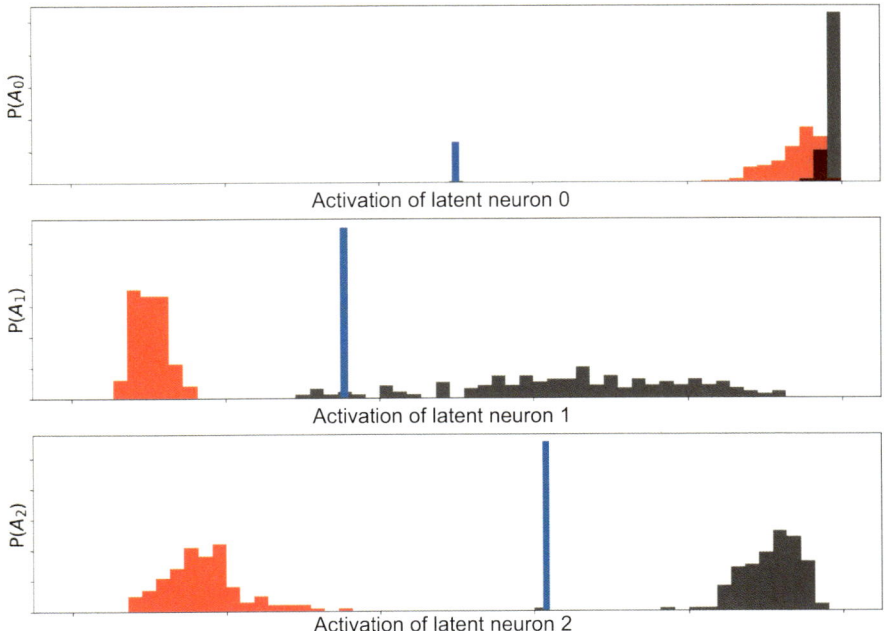

Fig. 5.11 Probability distributions obtained from the histograms for the activation of the latent neurons

But what about $P(A_2|A_1)$ or in our case with the specific numerical values $P(A_2 \approx 0.63|A_1 \approx 0.35)$? For this case too, we can create a histogram, we just need to filter the results beforehand and apply the condition $A_1 \approx 0.35$. We demonstrate this in Listing 5.16, where a conditional subset of events is filtered out.

Listing 5.16 Conditional probability in latent space

```
18   def conditionalSet(x, results, condNeuron1=1, condNeuron2=2,
         window=0.05):
19       mySet = []
20       for eachResult in results:
21           if x-window < eachResult[condNeuron1] < x + window:
22               mySet.append(eachResult[condNeuron2])
23       return np.array(mySet, dtype=float)
```

We use a window that is ±0.05 around the position of our neuron 1 with an activation of 0.35. From all cases that meet this condition, the activations of neuron 2 in mySet are sorted. They form a new set for which we can create a histogram.

Fig. 5.12 shows the result. If an activation at neuron 1 is approximately 0.35, then the activations for neuron 2 are all above 0.6. Thus, we can further secure our special case, because the histogram shows us that the probability for a normal case is vanishingly small.

Detection of an anomaly through the comparison of input and reconstruction

The second way to detect an anomaly is the quality of reconstruction. To investigate this, we can run through all test cases in our example and successively

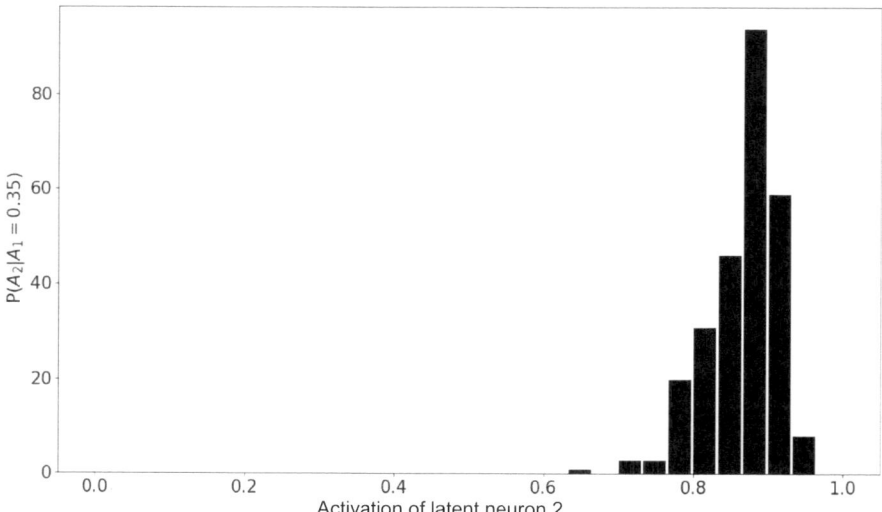

Fig. 5.12 Histogram of the conditional set, as a measure for the conditional probability for a latent neuron

calculate the correlation between input x and output z (the latent layer was y, behind the encoder, z is the result behind the decoder). This correlation is shown in Fig. 5.13 and is calculated in the following code example:

Listing 5.17 Reconstruction quality

```
18  reconstructionQuality = []
19  for i in range(0,20):
20      reconstructionQuality.append(np.corrcoef([Xtest[i],
            decodedGauss[i]])[0,1])
```

5.5.8 Combination of Methods

Autoencoders can be combined with other methods. To do this, the output from the latent layer is forwarded. Fig. 5.14 shows this schematically. Here, any subsequent methods can be used, the image shows a decision tree, as well as a classification or regression network as examples. In all the above example applications, the classification network is a solution to, for example, realize the assignment of the Gaussian functions to their classes.

The data flow concept shown in Fig. 5.14 also allows simultaneous operation of both subsequent methods. The results of the decision tree and neural network are then evaluated together.

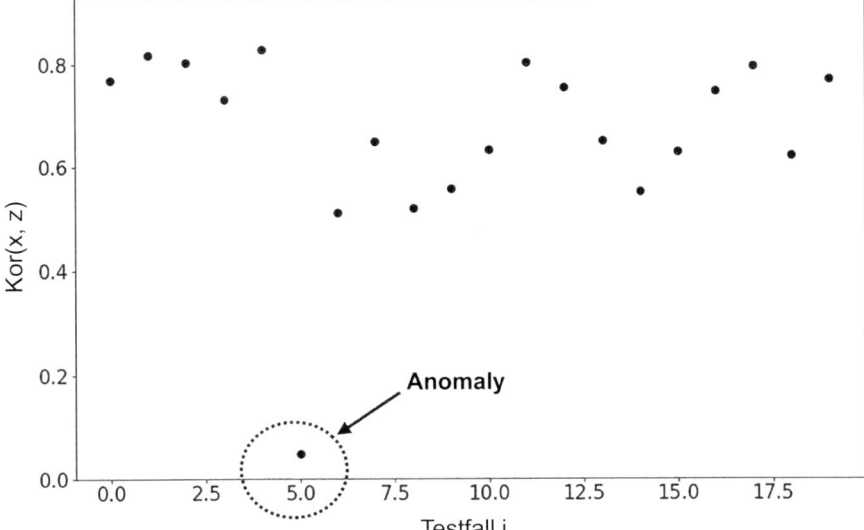

Fig. 5.13 Reconstruction quality per test case, outlier indicates the anomaly

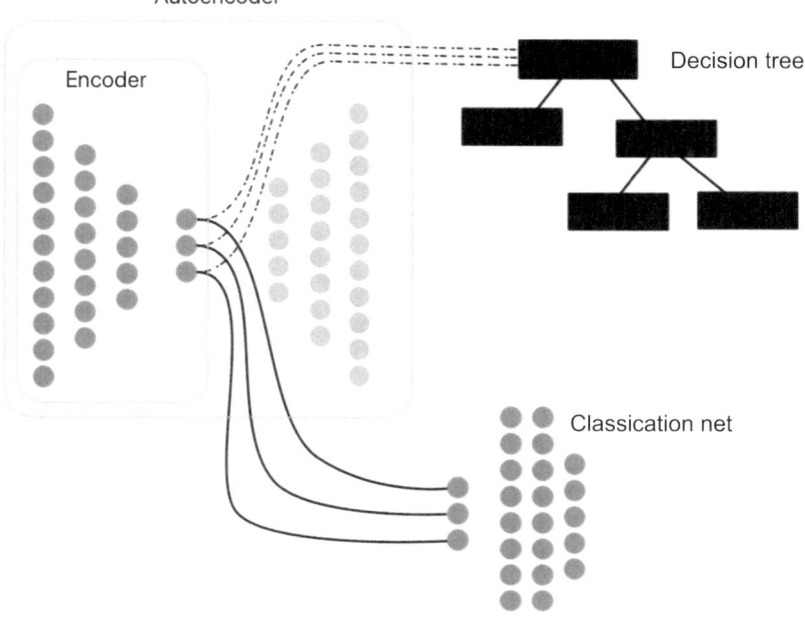

Fig. 5.14 Multi-stage combination of AE with a classification or regression network and with a decision tree

Stacking of Autoencoders
In a similar way, multiple autoencoders are connected to each other. This is also referred to as stacking of AE. The individual AE from a stack are trained individually, each on their own.

Summary
In this chapter, we have learned about learning methods that analyze the structure in data. They find groups of related data, clusters, which show similar behavior in processes in terms of their characteristics.

The K-Means method tries to determine cluster centers by geometrically shifting centroids. The aim here is to minimize the cluster variance.

The PCA puts the covariance matrix of the input data in the foreground. It analyzes the eigenspace of this matrix and allows irrelevant eigenvectors to be sorted out, thus reducing the dimension. Anomalies can also be detected in the eigenspace, an application that plays an important role in many real machines and systems.

t-SNE projects high-dimensional input data onto a low-dimensional space, often onto a plane. It ensures that the original distances are maintained as much as possible, while the individual clusters are separated.

Finally, we got to know the autoencoder, a neural network that maps the input variables to the output. The AE thus lays the foundation for classification problems and is a popular preprocessing stage for various subsequent methods. In its inner layer, the AE realizes a nonlinear compression and thus distinguishes itself from the PCA, a purely linear method.

A recurring concept is dimension reduction. Both the PCA and the autoencoder are capable of effectively reducing the dimension of the input data. Thus, large amounts of data are reduced to an optimized information content. For this reason, these two unsupervised learning methods are often found in the field of deep learning as preprocessing.

Tasks

5.1 Apply the PCA as described to the motor current example and feed the transformed data into the classification network from Sect. 4.5.7. What do you observe with regard to training times and prediction quality?

5.2 Create an autoencoder for the motor current example and train it on a suitable training set. Now add a classification network as in Sect. 4.5.7 and train it on the results of the autoencoder. Then compare your results with the result from task 5.1. What differences do you see?

5.3 Use the t-SNE application on the motor current example and train a classification network with the t-SNE results as described in Sect. 4.5.7 and in task 5.1 and 5.2. What causes the failure of the applicability of this direct approach?

5.4 Perform the batch K-means procedure on our motor current example. Specify a) 2 clusters, b) 4 clusters, c) 6 clusters, and d) 8 clusters for training. What differences do you see in the application?

References

1. D. Boardman and A. Flynn, "Performance of a fisher linear discriminant analysis gamma-ray identification algorithm," *IEEE Trans. Nucl. Sci.*, vol. 60, no. 2, pp. 482–489, 2013.
2. D. Boardman, M. Reinhard, and A. Flynn, "Principal component analysis of gamma-ray spectra for radiation portal monitors," *IEEE Trans. Nucl. Sci.*, vol. 59, no. 1, pp. 154–160, 2012.
3. G. Hinton and S. Roweis, "Stochastic neighbor embedding," *Advances in neural information processing systems*, vol. 15, pp. 833–840, 2003. [Online]. Available: http://citeseerx.ist.psu.edu/viewdoc/download?doi=10.1.1.13.7959&rep=rep1&type=pdf.
4. X. Jin and J. Han, *K-Means Clustering*. Boston, MA: Springer US, 2010, pp. 563–564. [Online]. Available: https://doi.org/10.1007/978-0-387-30164-8_425.
5. U. S. Kameswari and I. R. Babu, "Sensor data analysis and anomaly detection using predictive analytics for process industries," in *IEEE Workshop on Computational Intelligence: Applications and Future Directions*, Dec. 2015.
6. L. J. P. Maaten and G. E. Hinton, "Visualizing high-dimensionality data using t-sne," *Journal of Machine Learning Research*, vol. 9, pp. 2579–2605, Sep. 2008.

7. L. McInnes and J. Healy, "Umap: Uniform manifold approximation and projection for dimension reduction," *ArXiv*, vol. abs/1802.03426, 2018.

8. M. J. Neuer, *Quantifying Uncertainty in Physics-Informed Variational Autoencoders for Anomaly Detection*. Springer Nature, 2021.

9. M. J. Neuer, A. Quick, T. George, and N. Link, "Anomaly and causality analysis in process data streams using machine learning with specialized eigenspace topologies," in *Proceedings of ESTAD 2019*, 2019.

10. J. Newling and F. Fleuret, "Nested mini-batch k-means," 2016.

11. M. L. Shyu, S. C. Chen, K. Sarinnapakorn, and L. Chang, "Principal component-based anomaly detection scheme," *Studies in Computational Intelligence*, vol. 9, p. 19, 2006.

Chapter 6
Physics-Informed Learning

Keywords Physics-informed learning · Data enrichment · Differential equations · Integration · Automatic differentiation · Process corridors

A learning process can be supported by supplying a-priori known information, which is an approach called physics-informed learning. Within this concept, we demonstrate methods of data enrichment and the embedding of analytical expressions in neural networks. Afterwards, ways are outlined on how uncertainty can be integrated as an element into the learning processes. The focus is on the practical aspects necessary to apply such approaches in an industrial environment. The chapter ends with a look at the analysis of process corridors.

Adding already existing information reduces the training effort. Fewer data are needed, training times are shortened, and the result becomes interpretable. In this way, this chapter also affects all previous concepts, as outlined in Fig. 6.1. In addition, the methods presented here contribute to explainability.

6.1   Introduction

In the preceding chapters, we have learned how powerful unsupervised and supervised machine learning methods can be applied to production processes. Methods such as decision trees and neural networks can be used for classification and regression problems. Both are already of great value for industrial applications. Robust models are the heart of many control techniques, such as model predictive control or control with internal models. In optimization problems, predicting the future state of a system with a prediction model is also crucial.

M. J. Neuer, *Machine Learning for Engineers*,
https://doi.org/10.1007/978-3-662-69995-9_6

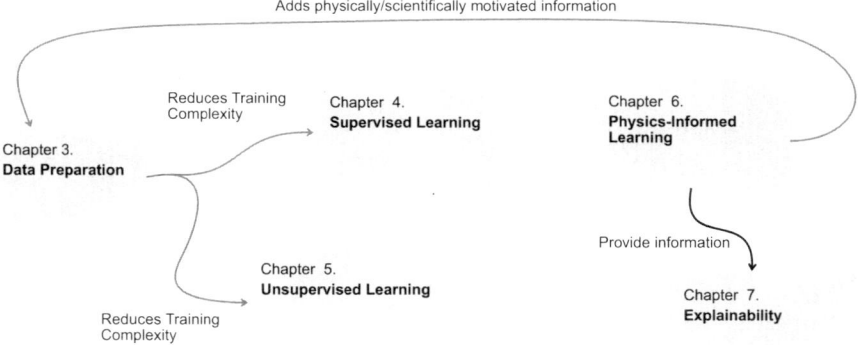

Fig. 6.1 Impact of Chap. 6 on the other chapters

In this chapter, we will focus on deliberately designing the learning algorithm to include certain pre-information, pursuing the question: „Why should we hide information from an algorithm when this information would actually improve its training and performance?" The answer to this question is not obvious and we should wisely weigh which information we can access at what time.

In the last chapters, we also established our understanding of physical measurements as stochastic processes. The measurement value $x(t)$ and the timestamp t are considered as quantities with uncertainties: $x(t) = \xi(t) + \delta\xi$ and $t = \tau + \delta\tau$. These uncertainties affect all technical process data and their consideration is of great interest for many technical solutions.

6.1.1 What is Physics-Informed Learning?

Physics-informed machine learning not only uses raw data as input for a machine learning algorithm but also provides additional, helpful information—*prior knowledge*—to the algorithm. From our point of view, prior knowledge can include transformations of the input data, functional dependencies, or differential equations—in general, any kind of law that helps the algorithm in training the model.

Many works limit the use of the term „physics-informed" to neural networks where physical laws are embedded in the architecture, often by enforcing an adapted loss function. We will take a broader view. For us, physics-informed machine learning refers to all those learning techniques that use prior knowledge in their approach. This includes, in particular, suitable mathematical transformations of the input data, if they are based on a consistent scientific law and are used to reduce the training effort. However, we will of course also discuss the embedding of equations and differential equations in network architectures in Sect. 6.3, as this is one of the most important methods.

Learning algorithms can indeed find the best transformations themselves. In particular, deep learning has proven to be very powerful for this task, with convolutional networks that can generate any number of nonlinear filter stages. Representation learning and transformation learning, as presented by [8], are also good examples of modeling even complex system behavior.

Why should we then bother to include prior knowledge? Because by providing the right transformations ab initio, these relevant pre-processings of the data do not have to be trained and the entire learning process becomes simpler. Of course, a neural network could learn the optimal representation from the dataset, but if we support it with a good transformation, this will reduce the network's training effort.

6.1.2 A Simple, Motivating Example

The effects of including prior knowledge can be better understood using the following simplified example:

Example: Motor Current with Physically-Informed Network
We examine some data from a motor. This data is given in the form of the measurement size μ—it is secondary what size it is. The motor has three *normal* operating modes, for which the curves μ are sorted into a data set **G** that represents the *good* cases. In some rare cases, the motor exhibits a certain failure mode and the corresponding curves for μ are collected in the data set **B** for *bad* cases. The distinction of these cases is a canonical classification problem. ◄

In Fig. 6.2, some example curves of μ are recorded and the different classes can be recognized. The green curves (and indeed there are several green curves) represent the bad cases. Obviously, we can visually recognize that the bad cases are characterized by an oscillation, even without having details about the process.

You can now certainly train a deep neural network that is large enough to internally form any nonlinear transformation that could find a specific solution leading to a good separation between good and bad cases. However, since the failure mode clearly exhibits these characteristic oscillations in some way, the underlying dynamics in the Fourier space are much easier to identify with the FFT method introduced in 2.4.

The Fourier Transformation (FT) is therefore well suited for this case as an enrichment for the algorithmic execution—it improves the learning process. Fig. 6.3a shows on the left side a regular classification network and Fig. 6.3b a similar network, where the FT is only added in addition to the original raw input data. This network is already a simple physics-informed neural network (PINN).

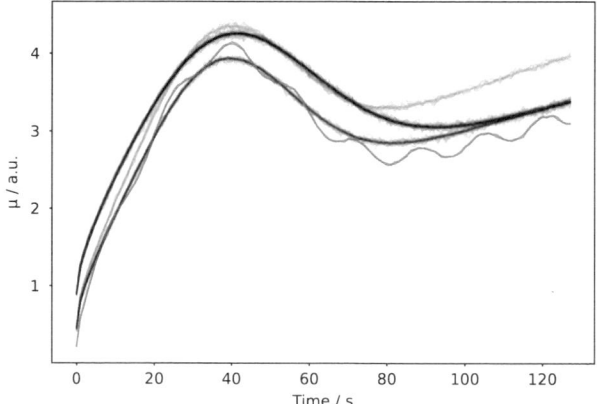

Fig. 6.2 Division of the motor current example into good and bad cases. The black curves belong to the „good" data set **G** and the green curves to the „bad" data set **B**

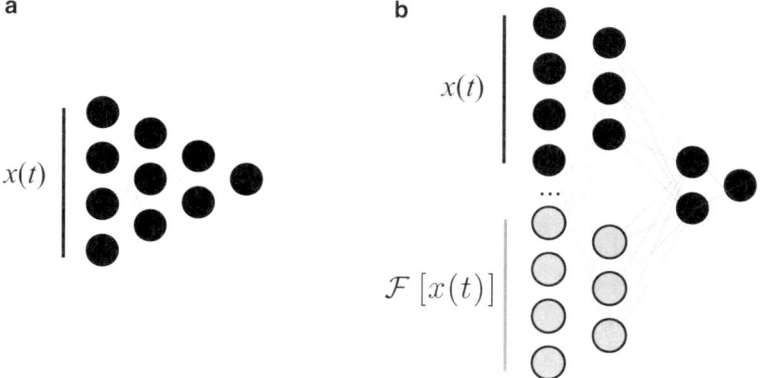

Fig. 6.3 Difference between classic network (left) and FFT-enriched, physics-informed network (right)

Note that there may also be situations where providing only the transformed (enriched) parts of the data is sufficient and the raw data can even be neglected. In fact, using the Fourier Transformation and focusing on the relevant frequency values would significantly reduce the amount of input data for classification.

One difficulty with this technique is the availability of *prior knowledge*. In the example, we saw the advantage of using a Fourier transformation based on the data itself. In other words, we needed an expert with process knowledge to tell us that including the Fourier transformation is a good idea. We are again involving human intelligence, rather than relying entirely on machine training. Later, we will see that these derivations can also be performed automatically with the help of a database for expert knowledge.

This inclusion of expert knowledge also leads to one of the most striking advantages of physics-informed approaches: The explainability of the form. If we train a network like the one on the left in Fig. 6.3, this network could correctly classify the raw time series. However, it does not give us any useful information about why the separation works so well. If we instead use a network like the one on the right side of Fig. 6.3, we can specifically look for the most influential input data. In the above example, the greatest influence would come from the Fourier-transformed input data, leading to the simple interpretation that an oscillation is the cause of the problem. The solution is now somewhat better explainable.

6.1.3 Continuous Change in the Process Chain

When implementing machine learning in a technical context, especially in the process industry, one obstacle to consider is the availability of sufficient training data. Production lines are not static. They require tools and equipment that are subject to constant wear and tear. Components and tools are replaced iteratively, which is a normal consequence of their use. In this sense, the production processes that generate the data we use for training change over time.

The modernization of processes and equipment is another reason for changes in production. Sometimes the processes evolve over time to improve product quality and process stability. The data of these aggregates, collected at different times, e.g., over two to three years, will change significantly.

The difficulties associated with machine learning solutions could be mitigated if there was a well-established, digital documentation of production changes. A first stage would be a logbook of tool changes, containing some ID numbers for the tools. This also represents a kind of data enrichment, realized here through an additional database.

6.1.4 Statistical Balance

The difficulty of an insufficient data basis is even exacerbated when assuming nearly perfect production processes. Faulty products may only occur rarely, in the worst case at a frequency lower than the change of processes, as described in Sect. 6.1.3. But even if no changes occur at all, it becomes a difficult task to find balanced amounts of good and bad cases (Fig. 6.4).

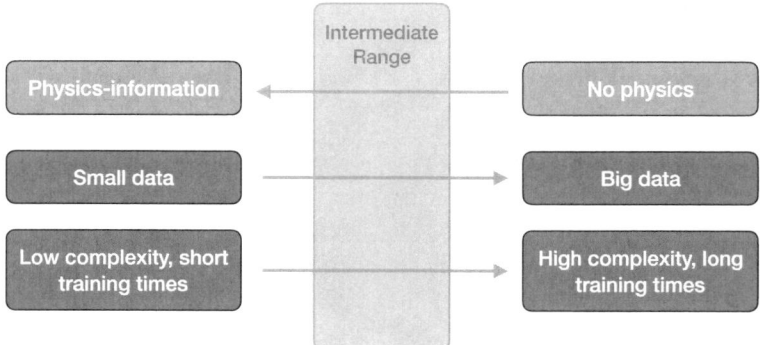

Fig. 6.4 Illustration of the relationship between the degree of physics to be included and the amount of data, based on the classification given in [6]. The complexity of the learning algorithm and the training times can also be reduced by adding external knowledge

6.1.5   Reasons for Physics-Informed Machine Learning

Machine learning can train various and very diverse types of relationships with the right input data. Of course, the input data must reflect the relationship in some way—if the necessary information is not contained in the data, there is no chance of actually learning it. However, once the dependency is contained in the data, the training will be successful if a) enough data is available and b) the hyperparameters of the learning algorithm are chosen correctly. But both Sects. 6.1.3 and 6.1.4 show that it is desirable, sometimes even necessary, to work with small amounts of data—one reason for using physics-informed approaches is therefore to apply machine learning to these significantly small data.

Another reason is that the inclusion of prior knowledge also helps us better understand the decision-making process within the learning algorithm. We can see this by revisiting our simple example from 6.1.2. Suppose you have provided the learning algorithm with both the raw data and the Fourier data and successfully trained it to predict good and bad product conditions. Then the next simple step would be to analyze which of the two vectors of input variables had the greatest influence on the result—a technique we will get to know as sensitivity analysis in the coming chapter. If this analysis shows that the FFT preparation has a significantly greater influence on the overall result, one can conclude that vibrations are responsible.

In other words: Physical prior information can be used in two different ways: First, they can reduce the training effort and help us deal with small amounts of data. Second, they also help in identifying those inputs that have the greatest influence on the result.

6.2   Data Enrichment

Data preparation in the sense of searching for the best representation of the data
has always been a part of data mining, which goes beyond the rudimentary tasks
of data cleaning and the exclusion of malicious data. However, this process was
rather limited to a pure adaptation of the data.

In physically-informed data enrichment, we explicitly refer to an active design
of the input data, in which we provide additional information to the machine learning algorithm. Active here means that we deduce from physical relationships and
considerations which additional information makes sense for the training. In this
form, either a conscious, cognitive process of selection originates from the pro-
grammer of the learning method, or there exists a level of knowledge from which
a learning method can automatically query this information. If you compare this
with Fig. 6.3, such an enrichment also preceded through the additional Fourier
transformation.

As mathematically trivial and comprehensible as this may be, from a practical
point of view, this question raises new questions for our data collection: Where
and when do we calculate the data enrichment?

6.2.1   Optimized Choice of Preprocessing

Let's first check which types of preprocessing we can consider suitable in connection with physical information. Many of these preparatory steps have already been
discussed in Chap. 3. Now we want to highlight their usefulness for designing efficient processing pipelines:

- **Histograms.** The use of histograms of the input data is a simple but effective
  measure to reduce the amount of data. A histogram reflects the probability distribution of the time series. In this sense, $x(t)$ is transformed into $P[x(t)]$
- **Derivatives.** As we have shown in Chap. 3, derivatives can help to sharpen the
  focus on certain properties. They remove constant parts from the input data and
  highlight changes.
- **Function curves.** If you suspect that a functional dependency is suitable
  as input, use it as a transformation. Many technical problems are subject to
  known physical laws, so we should simply prescribe these laws to the learning
  algorithm.
- **Shape properties (Features).** Identifying certain features in the input data
  and limiting learning to these features is also helpful, provided these features
  actually represent a good representation. As already mentioned, this task can
  of course also be effectively solved by representation learning—however, this
  requires a sufficient number of data points. If you can make a manual preselec-
  tion of significant features, this is a good choice for input data.

- **Frequency spectra.** In many situations, we are dealing with oscillating problems and here both the Fourier transformation and the various variants of the wavelet transformation are suitable transformations to shed more light on the details in the raw data.
- **Principal component analysis or Karhunen-Loevé transformation.** As one of the best techniques to reduce the dimensions to a minimum, PCA is always a sensible candidate for preparing your input data. One could argue that this does not directly correspond to the term *physics-informed*, and this criticism is justified. But one is not limited to performing the eigen space analysis only on the covariance matrix, one can also extend this idea to the eigen space e.g. of a wavelet extension.

6.2.2 Expert Knowledge on Data Objects

The use of physical information means that we must get the information from somewhere. There are several ways to achieve this.

- **(A) Data table**. The simplest approach is to store a table that links the variables and their respective transformation properties as shown in Fig. 6.5. This approach is certainly effective and one of the solutions that requires the least storage space. However, when working with digital twins, this method has the disadvantage that this separate table must be supplied with the data, otherwise important parts of the interpretation cannot be automated.
- **(B)** Meta-information **for each variable**. Another, somewhat more detailed option would be to provide the enrichment information with each stored data variable, as shown in Fig. 6.5. This has the advantage that only meta-information (short strings) need to be stored in the data object. Compared to (A), it is

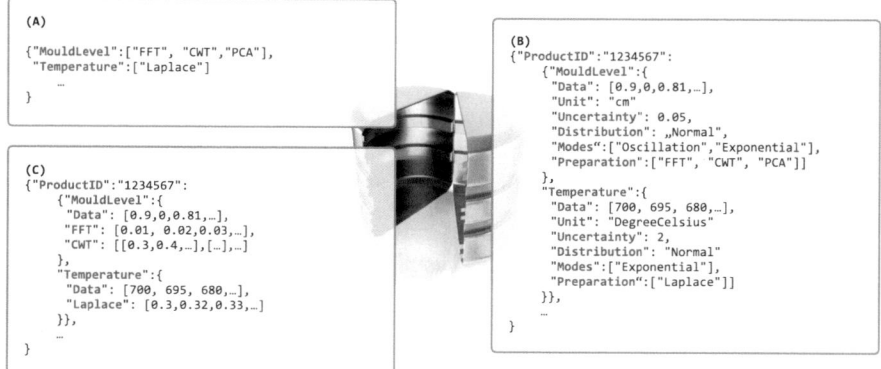

Fig. 6.5 Extension of data objects to consider ideal preprocessing steps for physically-informed machine learning

more storage-intensive, as each object stores redundant information. But now we can simply copy subsets of the object data to different places without needing an additional table and without losing the expert information.

- **(C) Fully enriched data.** The most storage-intensive solution is the calculation of data enrichment during the storage of the data or in an automatically triggered subprocess. All data is then included in the original data object that belongs to a product or machine. Although this option seems extreme, it is still the preferred approach to accelerate subsequent processes: The enrichment does not have to be calculated at training time or for visualization. The calculation was performed before these steps and thus this storage-intensive solution is also the fastest solution for further processing.

6.2.3 Automatic Optimization of Data Enrichment in Binary Classification

Let's now try a simple brute-force approach that helps us check the selection of suitable preprocessing steps. We focus on a case where we want to separate good and bad cases as well as possible, and therefore limit ourselves to a physics-informed optimization of data enrichment for binary classification. Let's assume that $\hat{\mathbf{O}}$ represents a vector with the transformation candidates, which are represented here by a corresponding operator,

$$\hat{\mathbf{O}} = \left[\frac{d}{dt}, \frac{d^2}{dt^2}, \exp, \log, \mathcal{F}, \mathcal{L}, \dots \right]. \tag{6.1}$$

In this sense, one can define any differential or integral operators and include them in (6.1). Once we have such a selection of transformations, we can construct a scalar product of all operator entries in $\hat{\mathbf{O}}$ with a weight vector $\boldsymbol{\omega}$,

$$\hat{\mathbf{L}}(\boldsymbol{\omega}) = \boldsymbol{\omega}\hat{\mathbf{O}} = \sum_{i=0}^{N} \omega_i \hat{O}_i, \tag{6.2}$$

where $\hat{\mathbf{L}}$ is really nothing complicated, but just a combination of the operators in $\hat{\mathbf{O}}$, each weighted with a factor.ω_i. Note that we choose ω as a symbol here to avoid confusing these weights with the weights w of a neural network. This combination of operators can act on $x(t)$. With the training data in the sets $\mathbf{G}$ and $\mathbf{B}$ we can find the ideal weights by solving the following optimization problem,

$$\underset{w_i}{\text{maximize}} \quad J_{\text{phys}} = \sum_k |\boldsymbol{\xi}_k - \boldsymbol{\zeta}_k|$$

$$\text{u. d. B.} \quad \text{(C1)} \ \boldsymbol{\xi}_k = \hat{\mathbf{L}}(\boldsymbol{\omega}) \boldsymbol{x}_k(t), \quad \text{with} \ x_k \in \mathbf{G},$$

$$\text{(C2)} \ \boldsymbol{\zeta}_k = \hat{\mathbf{L}}(\boldsymbol{\omega}) \boldsymbol{z}_k(t), \quad \text{with} \ z_k \in \mathbf{B}, \tag{6.3}$$

$$\text{(C3)} \ \omega_i \geq 0.$$

This optimization problem attempts to maximize the distance between ξ_k and ζ_k obtained by applying $\hat{L}$ to the example cases from **G** and **B**.

6.3 Embedding of Analytical Expressions in Neural Networks

Physics-informed learning also includes techniques to integrate differential equations (DE) into the network. Since the training of networks is based on an optimizer that performs a gradient descent, these gradients can also be used to find the solution of an ordinary DE.

In the work of Pfau et al. [11], this is demonstrated for the case of the Schrödinger equation, an important basic equation of quantum mechanics. Embedding of differential equations is also possible and helpful in recurrent networks, as the work of Nascimento et al. [9] shows. Some of the most fundamental works on this topic come from Karniadakis et al., with [6] being just one example.

6.3.1 Automatic Differentiation

Differentiation is by far one of the most common mathematical operations. It is the formal description of how one variable changes in relation to another, and thus a fundamental starting point for modeling dynamic systems with differential equations. The derivative $x'(t)$ of a function $x(t)$ can be calculated in three ways:

1. **Analytical Calculation**. Either with a classic pen-and-paper approach or with software that performs a symbolic evaluation, the derivative is determined according to the rules of analysis.
2. **Numerical Differentiation**. The data series $x(t)$ can be numerically derived to $x'(t)$, using the formula for the differential quotient or symmetric variants.
3. **Automatic Differentiation (engl. automatic differentiation)**. Here we use both the data and the functions stored in our optimization libraries. If we record the gradient descent method during optimization, we can always view the derivative.

In the next program examples, we will deal with ways to represent automatic differentiation in code. Libraries like Tensorflow offer practical tools for accessing the gradient of optimization.

In Listing 6.1, a simple derivative is shown. Here, Tensorflow's internal differentiation is applied to the function x**2. You can visualize the result from line 8 with Matplotlib.

Listing 6.1 Derivative using gradient tape

```
1  import tensorflow as tf
2
3  x = tf.Variable(tf.range(0,10, dtype=tf.float32))
4
5  with tf.GradientTape() as tape:
6      y = x**2
7
8  dy_dx = tape.gradient(y, x)
```

Partial derivatives are also possible here. To do this, they set up a function of several variables e.g. x1 and x2 and use GradientTape as shown in Listing 6.2.

Listing 6.2 Partial derivative using gradient tape

```
1  x1 = tf.Variable(tf.range(0,10,0.1, dtype=tf.float32))
2  x2 = tf.Variable(tf.range(0,10,0.1, dtype=tf.float32))
3
4  with tf.GradientTape() as tape:
5      y = x1**2+tf.sin(5*x2)
6
7  [dy_dx1, dy_dx2] = tape.gradient(y, [x1,x2])
```

These simple examples help us understand the functioning of GradientTape. Next, we will use this approach to implement further analytical expressions with a network.

6.3.2   Incorporating a Function as an Optimization Objective

The most basic operation is to replicate a function with a neural network. This sounds very similar to what we have seen before when we designed a regressor network to model functional dependencies. In this earlier approach, we defined an input vector t with times and then evaluated the equation of interest $f(t_i)$. t_i then represented the training input and $f(t_i)$ was the training output.

We will now implement an analytical relationship directly into the training of a network—which is nothing more than the previous approach, only that we insert the target function at a different point. This can be achieved by modifying the loss function J of the network,

$$J = \sqrt{\sum_i (\mathcal{N}(t_i) - f(t_i))^2}, \tag{6.4}$$

where $\mathcal{N}(t)$ is the output of the network. For the implementation, we use the regressor listed in Listing 6.3. However, we need to program individualized training for this learning method so that we can insert the Gradient Tape approach.

Listing 6.3 Regressor for including a function

```
import numpy as np
from tensorflow import keras
from tensorflow.keras import layers, losses, metrics
from tensorflow.keras.models import Model
import tensorflow as tf

class NNRegressor(Model):
    def __init__(self, inputLayerLength):
        super(NNRegressor, self).__init__()
        self.inputLayerLength = inputLayerLength
        self.construct_layers()
        self.regressor = keras.Sequential(self.mylayers)
        self.compile(optimizer='sgd',
                     metrics=[metrics.MeanAbsoluteError()])
        self.optimizer.learning_rate = 1E-3

    def construct_layers(self):
        self.mylayers = []
        self.mylayers.append(layers.Input(self.
            inputLayerLength))
        self.mylayers.append(layers.Dense(30, activation='
            sigmoid', use_bias=True,bias_initializer=tf.random
            .normal, kernel_initializer=tf.random.normal))
        self.mylayers.append(layers.Dense(30, activation='
            sigmoid', use_bias=True))
        self.mylayers.append(layers.Dense(1, activation='relu'
            , use_bias=True))

    def call(self, x):
        fittedData = self.regressor(x)
        return fittedData
```

The training call is made in its own class myEvaluator. It receives the regressor via the variable NN. In line 20, we set the network output equal to a function f and thus force the network to approximate this function during training.

Listing 6.4 Training call using gradient tape

```
class myEvaluator():

    def __init__(self, NN, f0=1):
        self.NN = NN
        self.f0 = f0
        self.loss = []

    def g(self, t):
        with tf.GradientTape() as tape:
            result = t * self.NN(np.array([t])) + self.f0
        return result

    def f(self, t):
        return t**2

    def my_loss(self):
        integral = []
        dt = np.sqrt(np.finfo(np.float32).eps)
        for t in np.linspace(0,2,6):
            integral.append((self.g(t) - self.f(t))**2)
        return tf.reduce_sum(integral)

    def step(self):
        with tf.GradientTape() as tape:
            self.loss.append(self.my_loss())
            self.NN.gradients = tape.gradient(self.loss[-1],
                self.NN.trainable_variables)
            self.NN.optimizer.apply_gradients(zip(self.NN.
                gradients, self.NN.trainable_variables))

    def train(self, number_of_steps=1000):
        for i in range(number_of_steps):
            self.step()
            if i%30==0:
                print('{} | Loss = {}'.format(number_of_steps -
                    i, self.my_loss().numpy()))
```

Finally, we call the regressor as follows:

Listing 6.5 Running the training

```
myNeuralNetwork = NNRegressor(inputLayerLength=1)
myEvaluator = myEvaluator(NN=myNeuralNetwork, f0=0)
myEvaluator.train(number_of_steps=10000)
```

The neural network now contains the function that we have specified. It is important to understand that we often use this process only for a part of overarching networks.

6.3.3   Integration of a Function

While the mapping of a function in the neural network can also be realized much more simply, neural networks can also solve more complicated analytical problems. This was shown in 1998 by Lagaris et al. in [7]. The authors provide solutions for the integration of ordinary and partial differential equations. The approach begins with a generic differential equation

$$x' = f(x, t), \tag{6.5}$$

for which we want to find a solution x. Next, we define this x as the output of a neural network $\mathcal{N}(t)$, which is fed with the input t,

$$\mathcal{N}(t) \approx x(t). \tag{6.6}$$

This is of course equivalent to

$$\frac{d\mathcal{N}(t)}{dt} \approx f(x, t), \tag{6.7}$$

which we can use as a basis for defining a loss function for our neural network,

$$J = \sqrt{\sum_i \left(\frac{d\mathcal{N}(t_i)}{dt} - f(x, t) \right)^2}, \tag{6.8}$$

which explicitly connects the differential operation in $f(x, t)$ with the network $\mathcal{N}$. An important part in solving a differential equation is to find a solution that satisfies the boundary conditions. To ensure this, Lagaris et al. in [7] suggested the following substitution

$$g(t) = x_0 + t\mathcal{N}(t). \tag{6.9}$$

We can see how helpful this substitution is when we consider $g(t = 0) = x_0$, which immediately leads us to $dg(t = 0)/dt = 0$. With g we can rewrite (6.8) to

$$J = \sqrt{\sum_i \left(\frac{dg(t_i)}{dt} - f(x, t) \right)^2}. \tag{6.10}$$

Eq. (6.10) is the loss definition for the neural network. When gradient descent trains the neural network to reduce the loss, g converges to the solution of the differential equation. The advantage of this approach is that training the neural network using gradient descent automatically optimizes (6.10) and ultimately numerically solves the differential equation. Details on partial differential equations can be found in [12] or [6], which provide further details on more complex applications.

In modern machine learning frameworks like Tensorflow, the change of the loss function can be done quite conveniently—and that's exactly what we need to minimize Eq. (6.10). We have seen how `tf.GradientTape` works for automatic differentiation in Sect. 6.3.1.

Using the same mechanism, before we tackle the solution of a differential equation, we can first integrate a function $f(t)$. We start again with the class `NNRegressor` from our above consideration, as described in Listing 6.3. We modify this with regard to the activation functions and the topology of the network as follows:

Listing 6.6 Regression network

```
1  import matplotlib
2  import matplotlib.pyplot as plt
3  import numpy as np
4  from tensorflow import keras
5  from tensorflow.keras import layers, losses, metrics
6  from tensorflow.keras.models import Model
7  import tensorflow as tf
8
9  np.random.seed(123)
10
11 class NNRegressor(Model):
12     def __init__(self, inputLayerLength):
13         super(NNRegressor, self).__init__()
14         self.inputLayerLength = inputLayerLength
15         self.construct_layers()
16         self.regressor = keras.Sequential(self.mylayers)
17         self.compile(optimizer='sgd',
18                      metrics=[metrics.MeanAbsoluteError()])
19         self.optimizer.learning_rate = 0.001
20
21     def construct_layers(self):
22         self.mylayers = []
23         self.mylayers.append(layers.Input(self.
                inputLayerLength))
24         self.mylayers.append(layers.Dense(50, activation='relu
                ', use_bias=True,bias_initializer=tf.random.normal
                , kernel_initializer=tf.random.normal))
25         self.mylayers.append(layers.Dense(50, activation='relu
                ', use_bias=True))
26         self.mylayers.append(layers.Dense(1, activation='relu'
                , use_bias=True))
27
28     def call(self, x):
29         fittedData = self.regressor(x)
30         return fittedData
```

We have parameterized the network so that it uses a stochastic gradient descent optimization and a fixed architecture with [l, 50, 50, 1], where l is the number of input variables. In addition, the initialization and bias are individually set for each layer to be able to experiment with them later. Leaky-ReLU was chosen as the activation function for all layers.

The next code, Listing 6.7, provides a class example for an integrator. It combines what we know about automatic differentiation and the neural network.

Listing 6.7 Integrator based on neural network

```
1  class myIntegrator():
2
3      def __init__(self, NN, fOperator, f0=1):
4          self.NN = NN
5          self.f = f
6          self.f0 = f0
7          self.loss = []
8
9      def g(self, t):
10         with tf.GradientTape() as tape:
11             result = t * self.NN(np.array([t])) + self.f0
12         return result
13
14     def my_loss(self):
15         integral = []
16         dt = np.sqrt(np.finfo(np.float32).eps)
17         for t in np.linspace(0,1,10):
18             dNN = (self.g(t+dt)-self.g(t))/dt
19             integral.append((dNN - self.f(t))**2)
20         return tf.reduce_sum(tf.abs(integral))
21
22     def step(self):
23         with tf.GradientTape() as tape:
24             self.loss.append(self.my_loss())
25             self.NN.gradients = tape.gradient(self.loss[-1],
                   self.NN.trainable_variables)
26             self.NN.optimizer.apply_gradients(zip(self.NN.
                   gradients, self.NN.trainable_variables))
27
28     def train(self, number_of_steps=1000):
29         for i in range(number_of_steps):
30             self.step()
31             if i%30==0:
32                 print('{}_|_Loss_=_{}'.format(number_of_steps -
                       i, self.my_loss().numpy()))
```

The constructor of the class `myIntegrator` requires the neural network and a function operator `fOperator` as input. Note that in lines 9 to 12, the substitution proposed by Lagaris et al. (6.9) is implemented in the class. This is also the crucial point where we use the presented `GradientTape` for automatic differentiation. It records the evaluation of the function g over the neural network and therefore allows us to access the gradients in lines 23 to 25.

Now we can specify a function f that should be integrated by this algorithm. The associated operator `fOperator`, which is used in the constructor of `myIntegrator` in line 3, can be passed via a lambda expression, which is shown in Listing 6.8.

Listing 6.8 Running the training for the integrator in Listing 6.7

```
1  f = lambda t: 2*t
2  myNeural_network = NNRegressor(inputLayerLength=1)
3  myIntegrator = myIntegrator(NN=myNeuralNetwork, fOperator=f,
       f0=1)
4  myIntegrator.train(number_of_steps=1000)
```

In Listing 6.8, $x' = f(t)$ is defined, followed by the instantiation of the neural network and finally an instantiation of the integrator. By calling the train() function of the integrator, the optimization process of the neural network is started, which performs the minimization of Eq. (6.10) in mathematical terms.

The analytical solution is also given here in the code. The integration starts at `f0=1`, which explains the offset of the parabola. In Fig. 6.6, the solution of the above code is shown. It quickly converges to a solution that is close to the analytical solution. The curves were generated with the code implemented in Listing 6.9.

Listing 6.9 Test of the integrator and visualization of results

```
1  def analytical_solution(x):
2      return x**2 + 1
3
4  result = []
5  tv = np.linspace(0,1,10)
6  for t in tv:
7      result.append(float(my_integrator.g(t)))
8
9  S = analytical_solution(tv)
10
11 plt.plot(tv, S, 'k--', label="Original Function", linewidth=2)
12 plt.plot(tv, result, label="Neural_Net_Approximation",linewidth
       =3.0, color=[0.0,0.6,0.6])
13 plt.legend(loc=2, prop={'size': 20})
14 plt.tick_params('both', labelsize=16)
15 plt.ylabel('x(t)_/_ a.u.', fontsize=18)
16 plt.xlabel('t_/_ i.u.', fontsize=18)
17 plt.show()
```

6.3.4 Integration of a First-Order Ordinary Differential Equation

While the previous example shows the integration of a function, we will now solve a simple differential equation. As an example, we apply the approach of automatic differentiation to the simplest ordinary differential equation (ODE),

$$x' = f(x,t) = x(t) \Leftrightarrow x(t) = x_0 \exp(t) + C, \tag{6.11}$$

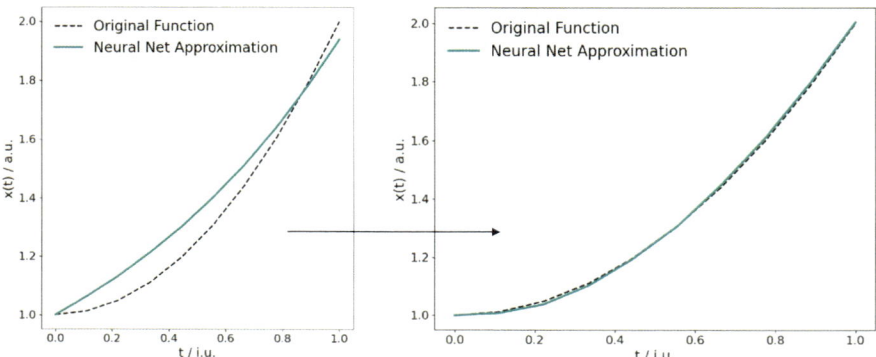

Fig. 6.6 Solutions for x in arbitrary units. Left: Intermediate stage of training, the loss is still greater than $L > 1.5$. Right: The solution of the integration has converged to the correct result, the loss is now $L < 0.02$

which is further simplified by assuming the boundary condition $x(0) = x_0$ and leaving $C = 0$. This simple ODE can be integrated with our previous code. The approach remains the same, as outlined by Lagaris et al. and already executed in the previous code examples. What we still lack is the representation of a differential operator, essentially the transition from a dependency only on t to a dependency on t and $x(t)$. Based on Listing 6.7, we remove the input of f via the lambda calculus in the code and instead add a function f to the integrator class. This is shown in Listing 6.10. It directly reflects the problem statement (6.11).

Listing 6.10 Changes to Listing 6.7

```
1      ...
2      def f(self,x):
3          return self.g(x)
4      ...
```

This expression alone does not yet correspond to the ODE. Without using the lambda expression, we can call the neural network to solve the ODE with the code in Listing 6.11,

Listing 6.11 Regression network for solving an ordinary differential equation

```
1  myNeuralNetwork = NNRegressor(inputLayerLength=1)
2  myIntegrator = myIntegrator(NN=myNeuralNetwork, f0=1)
3  myIntegrator.train(number_of_steps=1000)
```

which leads to the result shown in Fig. 6.7. Note that we have reused the test and result visualization code in Listing 6.9 to display the results. The integrator has found the expected exponential solution. The code can now be used as a starting point for any differential equations.

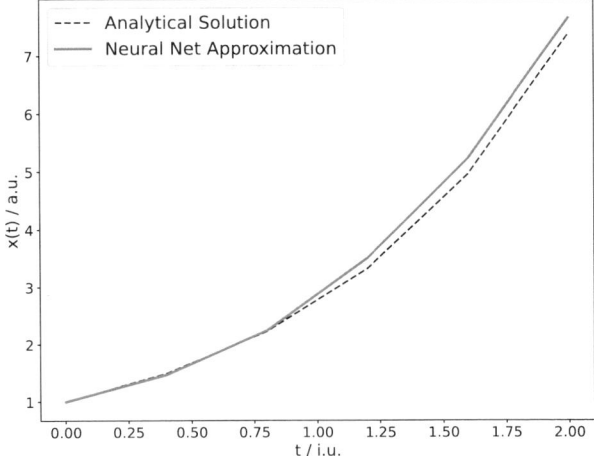

Fig. 6.7 Analytical solution $\sim \exp(t)$ (black) of the differential equation compared to the solution of the network (green)

6.4 Stochastic Methods for Integrating Uncertainty into the Learning Process

6.4.1 Considering Uncertainty in Input Data

Measurements are always associated with uncertainty. In Chap. 3 it was shown how the uncertainty in the measurement process itself arises and we were able to distinguish between aleatory and epistemic uncertainty. Since uncertainties are determined by stochastic processes, they can be captured with probability distributions. Many learning methods also use stochastic approaches. However, most of them with the intention to make processes faster and more efficient.

We now want to incorporate the uncertainty in our data more strongly into the methods. For this, it is extremely important to have this value—the uncertainty of a data point—available at all. As mentioned before, although industrial processes are now well digitized, knowledge of current measurement uncertainties is often not necessarily available.

A very simple but helpful strategy to incorporate uncertainties is the **stochastic enrichment**. This method increases the amount of data with a synthetic component, which however originates from the data itself and generates a representation of the uncertainty. The process of enrichment is as follows:

- **Selection of a subset of real data points.** The selection of a subset corresponds to the basic idea in training learning methods, which is to segment the data under consideration. We learned this as Batch K-Means in Chap. 6. To select the data points sensibly, we use a random draw that extracts an N-element subset from the M-element data point set.

- **Iteration over each data point of the subset.** We now move from point to point and determine its uncertainty. In the best case, there is an uncertainty per point in the data set. If the uncertainty is only given or even estimated via a one-time indication of the measurement accuracy, we assign exactly this value to each point. In some cases, the value must be calculated from the data set itself, using the variance as a helper.
- **Random draw from the uncertainty distribution.** We now have N data points with their uncertainties. Each data point has a probability distribution $p(\mu = x; \sigma = u_x)$ which represents the stochastic process. We choose the center of gravity of the distribution in such a way that it corresponds to the data point. When we draw new points from this distribution, we generate many points at its maximum (where the original measurement point was) and only a few points at places where the distribution is low. This new set of points is the **stochastically enriched data set**.

The outlined procedure raises some questions: Do we really have the uncertainties? Do we also know the probability density p? How many points do we need to generate to get a meaningful statistical selection? All these questions are strongly problem-related. How important they are is shown by current efforts in standardization processes, where it is demanded to capture the absolute or relative uncertainty for each data point and also the distribution p.

There are various approaches that are similar or even identical to stochastic enrichment. An example of this is the concept of the **kernel density estimator**. Here, too, a whole probability distribution is assigned to each point of a data group.

K-Means with stochastically enriched data
We now demonstrate stochastic enrichment using a simple example and again consider the use of the K-Means algorithm. The Batch K-Means algorithm introduced in Chap. 6 already works stochastically, in the sense that it makes a random choice in its process to reduce the amount of data points.

Now we apply the above stochastic consideration to the actual data. We already know that our data is burdened with uncertainties u_x. They therefore represent, each for themselves, the centers of individual probability distributions p. We now assume a Gaussian distribution of the data, which is often the best choice when a more precise indication of the expected distribution is missing. We also already know the function that realizes a random draw from the distribution from the K-Means example.

Listing 6.12 shows the random draw from a 2D Gaussian distribution, and how we generate additional points `stoch_x` and `stoch_y` from it.

Listing 6.12 Applying a stochastic K-Means method

```
def gauss2d(mu_x, sigma_x, mu_y, sigma_y, numberOfPoints=400):
    x1 = mu_x + sigma_x * np.random.randn(numberOfPoints)
    y1 = mu_y + sigma_y * np.random.randn(numberOfPoints)
    return x1, y1

XtrainWithUncertainty = []
stochasticEnrichment = 10
ux = 1.2
uy = 2.4
for i in range(0, 300):
    stoch_x, stoch_y = gauss2d(Xtrain[i][0], ux, Xtrain[i][1],
            uy, stochasticEnrichment)
    for j in range(0, stochasticEnrichment):
        XtrainWithUncertainty.append([stoch_x[j], stoch_y[j]])
```

These new data are assigned in the code to a new training variable XtrainWithUncertainty. This new training set can now be used in learning methods. Please pay special attention to where the uncertainties are used. In lines 8 and 9 we have defined them firmly to show you this in the example here. If you have specified individual uncertainties in your data set, ux[i] and uy[i] exist for you, and in line 11 of the example, Listing 6.13 must finally be used.

Listing 6.13 Stochasticity in the data

```
stoch_x, stoch_y = gauss2d(Xtrain[i][0], ux[i], Xtrain[i][1],
        uy[i], stochasticEnrichment)
```

To complete the example, Listing 6.14 shows the application of Scikit-Learn K-Means to the stochastically enriched dataset.

Listing 6.14 Applying Mini-Batch-K-Means to stochastically enriched data

```
miniBatchKmeans = MiniBatchKMeans(n_clusters = 3, batch_size =
        20, n_init = 10)
miniBatchKmeans.fit(XtrainWithUncertainty)
```

Discussion of the Method

What difference does the stochastic evaluation make compared to the original dataset? If you have no individual information about the uncertainty of each point, the added value of the method is low. The data points themselves must effectively also contain the probability distribution. In Fig. 6.8 both datasets are shown, the original points (black) and the enriched points. The strongest effect we see is in the uncertainty of the resulting cluster points. There are further applications where stochastic enrichment becomes important and brings about relevant differences in the interpretation of results.

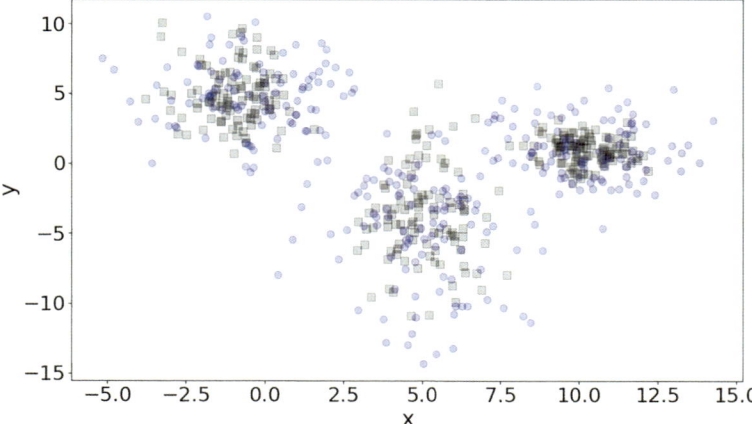

Fig. 6.8 Comparison of the original dataset (black) and the stochastically enriched dataset (blue)

One example is the temporal change in uncertainty. It can have a strong influence on the results: measuring devices can age, their resolution can deteriorate, or the physical process on which the measurement is based can undergo fluctuations. Imagine a measurement process based on a constant electrical voltage, where there is a high level of uncertainty for defined times. For such problems, stochastic enrichment is predestined. It can then achieve good representation for the respective sub-point.

6.4.2 Quantification of Uncertainty from the Learned Results

We want to determine now how the consideration of uncertainty affects the actually predicted results. For this purpose, we take one cluster center as an example, namely the cluster at the approximate position (5,−5). To determine the resulting uncertainty, we conduct several trials in succession and record the positions of the centers in an array. This procedure is comparable to Monte Carlo methods, only that our inputs are based on real data and our distributions on real uncertainties.

Listing 6.15 shows this approach. A loop runs through 500 trials. Each trial rolls a new configuration for the stochastic dataset and trains the K-Means method once. After the training, the centroids are stored.

Listing 6.15 Scanning through the stochasticity of the K-Means procedure

```
cluster1x = []
cluster1y = []

for trial in range(0,500):

    XtrainWithUncertainty = []
    stochasticEnrichment = 10

    for i in range(0, 300):
        ux, uy = gauss2d(Xtrain[i][0], 1.4, Xtrain[i][1], 2.2,
            stochasticEnrichment)
        for j in range(0,stochasticEnrichment):
            XtrainWithUncertainty.append([ux[j],uy[j]])

    miniBatchKmeans = MiniBatchKMeans(n_clusters = 3,
        batch_size = 20, n_init = 10)
    miniBatchKmeans.fit(XtrainWithUncertainty)

    for k in range(0,3):
        xc = miniBatchKmeans.cluster_centers_[k][0]
        yc = miniBatchKmeans.cluster_centers_[k][1]
        if xc > 2 and xc < 7:
            cluster1x.append(xc)
            cluster1y.append(yc)
```

To see how much the cluster center (of the selected cluster) varies, we can take a closer look at the histograms for the x and y coordinates. The code in Listing 6.16 shows the creation of the histogram.

Listing 6.16 Plotting the histograms for the clusters

```
fig, (ax1,ax2) = plt.subplots(2,1,figsize=(16,9))
ax1.hist(cluster1x,bins=20, rwidth=0.9)
ax1.set_xlim([3,8])
ax2.hist(cluster1y,bins=20,rwidth=0.9)
ax2.set_xlim([-3,-8])
```

Fig. 6.9 shows the histograms, where the result for our example is not surprising. The centroids of the histograms must necessarily lie at the original cluster center. We have defined the shape of the distribution and the width of the spread through our code.

The results shown here will be revisited later for the discussion of trust in learning methods. This approach works for all learning methods and was only carried out exemplarily for K-Means here.

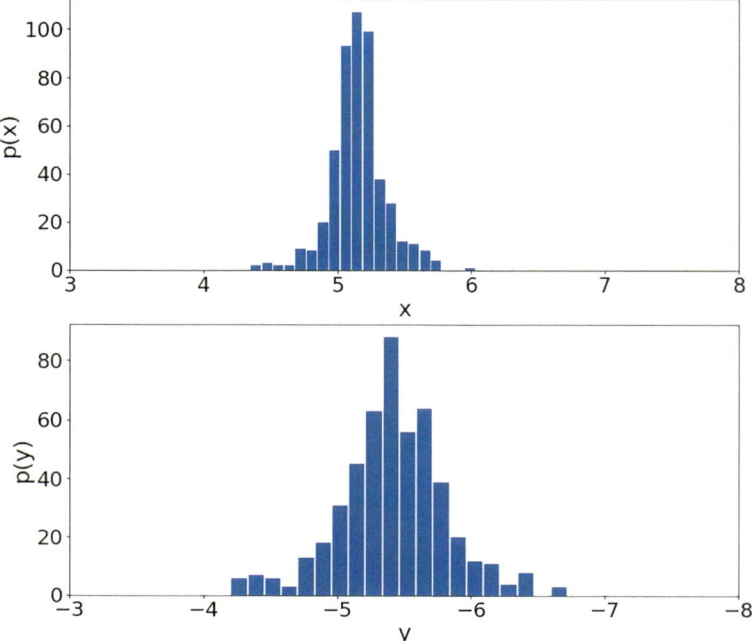

Fig. 6.9 Histograms of x and y coordinates of the cluster center

6.4.3 Stochastic Enrichment of Target Variables

Now that we know a way to integrate uncertainty into the evaluation and have practiced this for input data of an unsupervised learning algorithm, we now extend our consideration to fuzzy classes. They represent a case of uncertainty of the target variables. For this purpose, we revisit the classifier from Chap. 5, which we wrote in Keras. This classifier used a special assignment function that generates an image on classes. Its assignment was sharp in Chap. 5. A curve of the training or test data belonged to a clearly defined category, e.g., category 3, which was indicated by the vector [0, 0, 0, 1, 0].

We now imagine these target variables as a probability statement. Then it may be that the assignment to a category is no longer completely certain. The functional progressions of the motor current data are a good example of this. In Fig. 3.2 you can see that several curves are very close to each other and difficult to distinguish.

If you have access to stochastic labels, these can be taken into account in the classifier. As an example, in 6.17 a probabilistic assignment of the original labels to probability values was demonstrated for the motor current. These values can come from the measurement process or the automatic recording of labels. A scanner that scans a number and is only 95 % accurate, a color detection that maps intensity values to a color scale, or a correlation that can never be perfect, all these are examples of uncertain labels.

Listing 6.17 Stochastic labels

```
Xtrain = []
Ytrain = []
Xtest = []
Ytest = []

def probabilityForLabel(label):
    if label == 1:
        probabilityLabel = [0,0.84,0,0,0]
    elif label == 2:
        probabilityLabel = [0,0.1,0.5,0.1,0.1]
    elif label == 3:
        probabilityLabel = [0,0,0,0.6,0.1]
    else:
        probabilityLabel = [0,0,0,0,0.7]
    return probabilityLabel

for i in range(0,1400):
    Xtrain.append(X[i])
    Ytrain.append(probabilityForLabel(data['Label'][i]))

for i in range(1401,1420):
    Xtest.append(X[i])
    Ytest.append(probabilityForLabel(data['Label'][i]))
```

To train these labels, we need changes to the classifier. For one thing, the function `to_categorical` is obsolete, because we already cover this step with the probabilityForLabel() function in the above code.

The code in Listing 6.18 shows the classifier, with the few changes. The only things to highlight here are the adjustment of the cost function to MeanSquaredError() and the change in the Fit call, explicitly using `Ytrain` again. We train the classifier and finally test it with the Listing 6.19.

Listing 6.18 Classifier for stochastic labels

```
class Classifier(Model):

    def __init__(self, inputLayerLength, hiddenLayers=2):
        super(Classifier, self).__init__()
        self.inputLayerLength = inputLayerLength
        self.hiddenLayers = hiddenLayers
        self.constructLayers()
        self.classifier = tf.keras.Sequential(self.myLayers)
        self.compile(optimizer='adam',
                     loss=losses.MeanSquaredError())
        self.optimizer.learning_rate = 0.002

    def constructLayers(self):
        self.myLayers = []
        self.myLayers.append(layers.Input(self.
            inputLayerLength))
        for i in range(0,self.hiddenLayers):
            self.myLayers.append(layers.Dense(50, activation='
                leaky_relu'))
            self.myLayers.append(layers.Dense(50, activation='
                leaky_relu'))
        self.myLayers.append(layers.Dense(5, activation='
            softmax'))

    def call(self, x):
        classified = self.classifier(x)
        return classified

classifier = Classifier(len(Xtrain[0]))
history = classifier.fit(np.array(Xtrain),
                         np.array(Ytrain), epochs=150,verbose=
                             False, batch_size=50)
```

Listing 6.19 Classifier for stochastic labels

```
result = classifier.predict(np.array(Xtest))
for i in range(0,19):
    print('{}_|_{}'.format(np.round(result[i],2), Ytest[i]))
                         np.array(Ytrain), epochs=150,verbose=
                             False, batch_size=50)
```

This test leads to an output in the following form:

Listing 6.20 Testing the classifier for stochastic labels

```
...
[0.06 0.06 0.06 0.65 0.17] | [0, 0, 0, 0.6, 0.1]
[0.06 0.06 0.06 0.66 0.16] | [0, 0, 0, 0.6, 0.1]
[0.06 0.06 0.06 0.65 0.17] | [0, 0, 0, 0.6, 0.1]
[0.04 0.14 0.54 0.14 0.14] | [0, 0.1, 0.5, 0.1, 0.1]
[0.04 0.05 0.07 0.07 0.77] | [0, 0, 0, 0, 0.7]
[0.03 0.86 0.04 0.03 0.04] | [0, 0.84, 0, 0, 0]
[0.06 0.06 0.06 0.65 0.17] | [0, 0, 0, 0.6, 0.1]
...
```

The results that the network predicts, in the output on the left side, now reflect the probabilities of the labels.

6.4.4   Mixture-Density Networks (MDN)

Basis of MDN

A goal for every algorithm is the prediction of its own accuracy. This prediction is often difficult to determine in learning methods, because in learning methods it depends on the quality of the input data. Each individual measurement process can in principle be subject to different uncertainty. Thus, the uncertainty can be captured, as shown in [3], among others.

A Mixture-Density Network is based on the basic idea that not a fixed value is predicted, but rather a distribution function. They determine the parameters for this distribution in their hidden layers and allow us to access the average prediction and its uncertainty. In other words, Mixture-Density Networks can predict their own uncertainty. The most important work on this topic is the book by Bishop [1], who also goes into similar approaches in other publications [2, 4]. Some general approaches for Exponential Mixture Modelling can be found in Julier et al. [5].

We will explain the ideas here using a simple example, but we will not perform a complete derivation of the MDN. To understand the practical application, it helps to focus on an easy-to-understand figure $y = f(x)$ that we want to model with an MDN. As a result, we are interested in the probability distribution $p(y|x)$, i.e., how high the probability of a value y is, assuming that x has occurred.

This distribution is assumed in MDN as,

$$p(y|x) = \sum_{i}^{K-1} \omega_i \mathcal{P}[\mu_i(x), \sigma_i(x)], \qquad (6.12)$$

In total, we consider K different components, e.g., $K = 3$ Gaussian functions, as shown on the right side in Fig. 6.10. For each of these distributions, we need an amplitude or a mixing weight. This weighting is done via ω_i. Of course, this

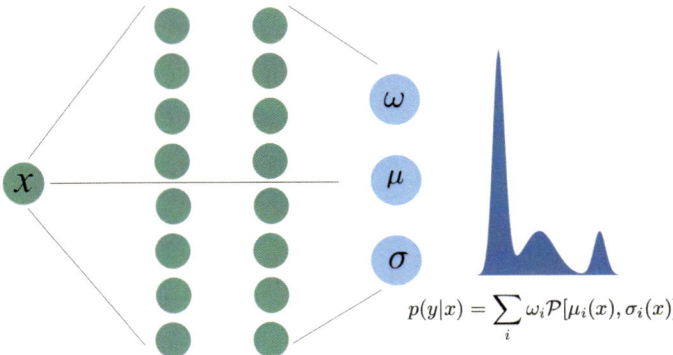

Fig. 6.10   Schematic representation of a mixture-density network

weight is part of the training and will be predicted by the model later. Ultimately, $\mathcal{P}[\mu(x), \sigma(x)]$ captures the distribution function itself, e.g., a Gaussian function,

$$\mathcal{P} = \exp\left(-\frac{(x - \mu)^2}{2\sigma^2}\right). \tag{6.13}$$

Forward calculation of an MDN

Important to the concept of the MDN is the „mixture" of several distributions. The mixture factors ω use a softmax activation,

$$\omega_i(x) = \text{softmax}(\boldsymbol{W}_\omega \boldsymbol{h}(x) + \boldsymbol{b}_\omega), \tag{6.14}$$

since the sum of all factors must be 1. For the means, we can use an activation like $f(x) = \text{ReLU}(x)$. However, we are not dependent on it, so we generally write the activation only with f,

$$\mu_i(x) = f(\boldsymbol{W}_\mu \boldsymbol{h}(x) + \boldsymbol{b}_\mu). \tag{6.15}$$

The width of a distribution is always greater than 0, and therefore we must select a special activation for the σ. It must not fall below zero. For this, we use the Exponential Linear Unit (ELU). It is defined as

$$\text{ELU}(x) = \begin{cases} \exp(x) - 1 & \text{für } x < 0, \\ x & \text{für } x \geq 0. \end{cases} \tag{6.16}$$

With this definition, σ is calculated in the last layer over

$$\sigma_i(x) = \text{ELU}(\boldsymbol{W}_\sigma \boldsymbol{h}(x) + \boldsymbol{b}_\sigma) + 1, \tag{6.17}$$

where the ELU function and the added offset ensure that Sigma only takes positive values.

Eq. (6.14) to (6.17) describe the forward evaluation of the network up to one last step, because in the output layer only parameters are output. A usable result is only obtained when a total distribution from the mixture of sub-distributions has been generated from these parameters. From this total distribution, a prediction is finally generated for a fixed x by a random draw.

Implementation of an MDN

To apply an MDN, we first create a suitable training and test set. For this, we choose a case that is as simple and understandable as possible,

$$y = x + x^2 * \exp(v), \tag{6.18}$$

where v is a random number between 0 and 1. Listing 6.21 shows the generation of these sample data.

Listing 6.21 Example data to demonstrate the use of a MDN

```
1   def generateSyntheticData(n=20):
2       xout = []
3       yout = []
4       for i in range(0,n):
5           x = 10*np.random.random(1)
6           y = x+0.05*x**2*np.exp(np.random.random(1))
7           xout.append(x)
8           yout.append(y)
9       return xout,yout
10
11  Xtrain, Ytrain = generateSyntheticData(n=2000)
12  Xtest, Ytest = generateSyntheticData(n=200)
```

We now turn to the actual MDN regressor. The code in Listing 6.22 builds on our previous Keras models, but includes some special features. In line 17, we use a special cost function for MDN. Also, in the construction of the layers, an MDN layer is inserted as the output layer. It comes from the „mdn" library which provides Keras with all necessary additional functions to handle MDN.

Listing 6.22 MDNRegressor using Keras

```
1   import keras
2   import numpy as np
3   import tensorflow as tf
4   from tensorflow.keras import layers, losses
5   from tensorflow.keras.models import Model
6   import mdn
7
8   class MDNRegressor(Model):
9
10      def __init__(self, inputLayerLength, hiddenLayers=2):
11          super(MDNRegressor, self).__init__()
12          self.inputLayerLength = inputLayerLength
13          self.hiddenLayers = hiddenLayers
14          self.constructLayers()
15          self.mdnRegressor = tf.keras.Sequential(self.myLayers)
16          self.compile(optimizer='adam',
17                       loss=mdn.get_mixture_loss_func(1,3))
18          self.optimizer.learning_rate = 0.001
19
20      def constructLayers(self):
21          self.myLayers = []
22          self.myLayers.append(layers.Input(self.
                inputLayerLength))
23          for i in range(0,self.hiddenLayers):
24              self.layers.append(layers.Dense(25, activation='
                  leaky_relu'))
25          self.myLayers.append(mdn.MDN(1, 3))
26
27      def call(self, x):
28          regressorResult = self.mdnRegressor(x)
29          return regressorResult
30
31  mdnRegressor = MDNRegressor(len(Xtrain[0]))
32  history = mdnRegressor.fit(np.array(Xtrain), np.array(Ytrain),
        epochs=300, batch_size=200)
```

In lines 17 and 25, specific details about the form and goals of the distribution are inserted. It is supposed to predict a variable and use 3 mixtures for this purpose. These are parameters that both the cost function and the MDN layer require.

Fig. 6.11 shows the comparison of test points (black) and the network prediction (red). The red crosses indicate the position of the prediction and the red error bars represent the uncertainty. The program code that tests the results of the network and creates the image is listed in Listing 6.23.

Listing 6.23 Plotting the test of the MDN

```
1   results = mdnRegressor.predict(np.array(Xtest))
2   y_samples = np.apply_along_axis(mdn.sample_from_output, 1,
        results, 1, 3, temp=1.0)
3   mus = np.apply_along_axis((lambda a: a[:3]), 1, results)
4   sigs = np.apply_along_axis((lambda a: a[3:3*2]), 1, results)
5
6   plt.scatter(Xtest,Ytest, color='k')
7   for i in range(0,len(y_samples)):
8       #plt.scatter(i, results[i][2], s=140, marker='x',linewidth
            =5,color='k')
9       plt.scatter(Xtest[i], y_samples[i], s=140, marker='x',
            linewidth=5,color='r',alpha=0.2)
10      plt.errorbar(Xtest[i], y_samples[i], yerr=sigs[i][1],color
            ='r',alpha=0.5)
11  plt.tick_params('both', labelsize=22)
12  plt.xlabel('x', fontsize=24)
13  plt.ylabel('Ytest[x], result[x]', fontsize=24)
```

Each execution of the network leads to a result `result`. This result, calculated in line 1 of the listing, only encodes the distribution. In order to be able to check the quality of the prediction, a sample must still be extracted from the distribution. This happens in line 2, where we execute `sample_from_output` to draw from the distribution.

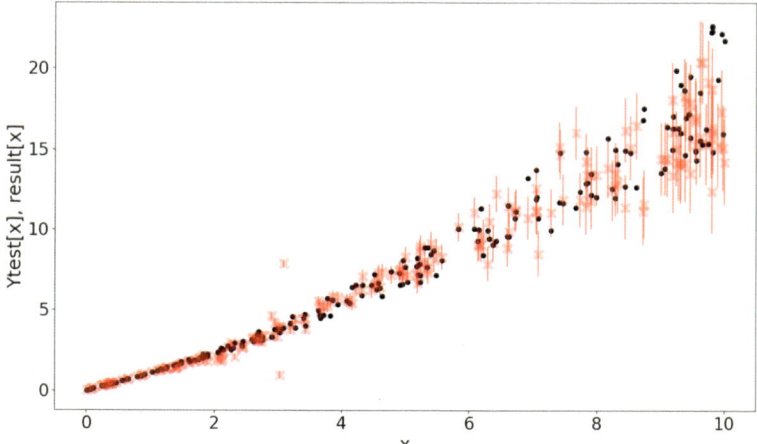

Fig. 6.11   Result of the evaluation of the MDN

6.5   Process Corridors

6.5.1   Probability Densities through 2D Histograms

As we have already seen in Sect. 3.7.1 have, histograms reflect the probability densities of data variables. If one uses a 2D-histogramming, one can visualize the probability corridor along a time series or several variables. If these probability distributions are based on a process chain, we also speak of a process corridor.

> **Process corridor.** A group of probability distributions $P_i(x, \pi)$ of process variables x and parameters π is called a process corridor if each distribution can be assigned to a process i.

We would like to explain this in more detail using an approach from [10].

Corridor for Time Series
Fig. 6.12a shows a series of normal process curves of any repeating process. We can assume that each measured data series should behave similarly to its predecessor. There are several known anomalous events in this signal, but the most difficult to detect effect is an oscillating disturbance of the voltage.

Formally, we define **G** as the set of all normal (good) situations and **B** as the set of all bad scenarios. For our example, we can of course further divide these sets into G_{Train}, G_{Test}, B_{Train} and B_{Test}.

We continue our considerations by taking a very simplified perspective. First, we want to correctly describe our uncertainty corridor along the signal. To do this, we let $x \in X$ be a stochastic process, namely a time series with discrete time steps, which describes a process variable of a production process. Then, $P(x|t_j)$ with $j \in \mathbb{N}$ describes the probability of finding the value x at a certain point in time t_j.

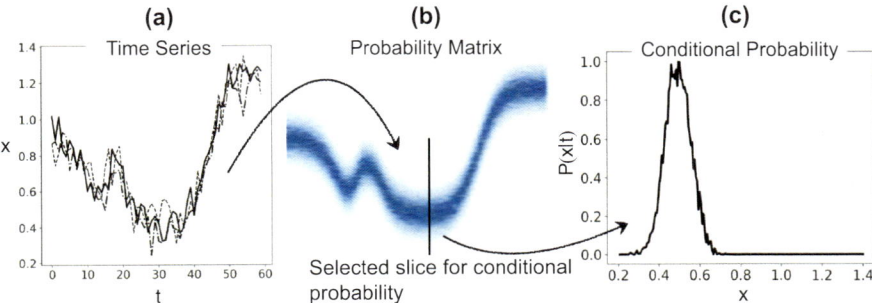

Fig. 6.12 Illustration of a probability corridor of the variable $x(t)$. Various measurements of $x(t)$ are shown in (**a**). (**b**) shows the resulting 2D histogram. Each cut of this histogram represents a conditional probability density, and (**c**) reproduces one of these cuts

The set X can be further discretized by a limited number of intervals $\xi_i = [\xi_{i_<}, \xi_{i_>}]$, so that with

$$\xi_{i_<} < x < \xi_{i_>} \Rightarrow x \in \xi_i \qquad (6.19)$$

a probability matrix $\mathbf{P} = (P_{i,j})$ can be defined,

$$P_{i,j} = P(x \in \xi_i | t_j). \qquad (6.20)$$

The size of $\mathbf{P}$ depends on the number of intervals ξ_i and the number of time steps t_j. $\mathbf{P}$ encompasses the entire analytical movement $\tilde{x}(t)$, more precisely, the path described by $\tilde{x}(t)$ is the distribution mean of $\mathbf{P}$.

Once (6.20) is calculated from the data, it is already a practical tool for detecting anomalies in the process. Every occurrence of x in an interval ξ_i with a low probability indicates an anomalous behavior of the process. Fig. 6.12 illustrates this by assuming a process time series in Fig. 6.12a. With a sufficiently large data set, the matrix $\mathbf{P}$ can be trained to reflect the statistical behavior of the time series, as shown in Fig. 6.12b, where the intensities scale with the numbers of the matrix. Finally, the column vectors of the matrix $\mathbf{P}$ at t_j naturally represent the individual conditional probability distributions.

The method shown or very similar variants of it are often found in industry for heuristic detection of anomalies. It only requires knowledge of some statistical parameters and an estimate of the shape of the normal process curve. Although this probability corridor is a good approach for observing deviations from the path, it has several weaknesses:

1. It does not recognize any deviations that remain within the statistically defined normal path, even if the main shape is significantly distorted. This is the case, for example, when a sinusoidal noise disturbs the original signal, whose amplitude is small enough to be captured by the noise fluctuation.
2. The assumption that the distribution $P(x|t_i)$ is independent of the previous position $x(t_{i-1})$ is a strong simplification. It implies that for each position $x(t_{i-1})$ the subsequent probability distribution for the next time step is the same.
3. Any continuous deterioration over time, which is used within the scope of this approach, is merely reflected in a widening of the probability corridor and thus does not adequately account for the presence of an anomaly.

Implementation of the probability corridor

There are many variants to use 2-histogramming to create a such corridor. We choose here our own code. It allows us to build on this basis later and extend this concept to conditional probabilities.

In Fig. 6.13, a grid is shown, which discretely samples the course of a function (black line). For this, we have created windows that are defined by their x- and y-positions. Whenever the black curve comes through a window, a counter is increased in it. At the beginning, all windows are set to 0. We now carry this out for each repeated time series. After several time series, the process corridor is then created as it is depicted in Fig. 6.12b.

Listing 6.24 shows how to program this method, using only Numpy and without the help of additional libraries, in Python. A discretization of the two abstract

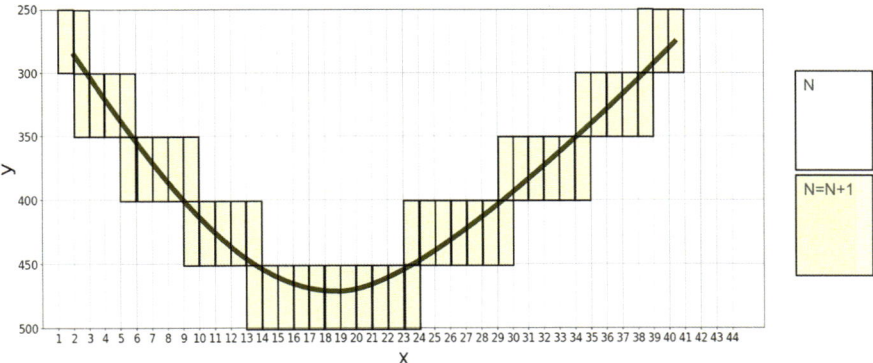

Fig. 6.13 2D histogram with count of events per element

axes is given, i.e., how many windows you want to consider in total. This corresponds to the bins in the normal, one-dimensional histogram, and therefore the individual windows in the code are referred to as bins.

Listing 6.24 Simple probability corridor using 2D histogramming

```
%matplotlib tk
import pickle
import numpy as np
import matplotlib.pyplot as plt

class probabilityCorridor(object):

    def __init__(self, data, x, y):
        self.x = x
        self.y = y
        self.bin = np.zeros([y+1, x+1]) #x+1,y+1])
        self.train(data)

    def train(self, data):
        dataMatrix = np.array(data)
        self.ymax = dataMatrix.max()
        self.ymin = dataMatrix.min()
        self.scalefactor = self.y / (self.ymax-self.ymin)

        for eachSet in data:
            for x in range(0,self.x):
                y =  int(np.floor(self.y *(eachSet[x]-self.
                    ymin)/(self.ymax-self.ymin)))
                self.bin[self.y-y][x] += 1

        return self.bin

dt = probabilityCorridor(data['X'][0:400],100,100)
dt.train(data['X'][0:400])
plt.imshow(dt.bin, vmin=0, vmax=20)
```

The training scales the data and increases a counter in the corresponding bins whenever the curve crosses the window. Windows with high counter values are therefore hit more frequently by the time series, those with low values less frequently.

Corridor for multiple sensor variables

A process corridor does not necessarily have to be applied to a time series. The method can also be applied to, for example, several different variables from sensors. It then makes sense to normalize and scale these beforehand—two very problem-dependent steps.

Such a corridor is shown in Fig. 6.14 for 26 individual sensors of a temperature measurement. As an example, a specific, anomalous measurement is shown in the form of points in the diagram. At one point, the measurement points leave the corridor.

Summary

In this chapter, we have discussed physics-informed, machine learning, where neural networks are supplied with prior knowledge to improve their prediction or to embed the prior knowledge in the algorithmic process.

We have seen that we can use the optimization engine of ML frameworks to perform automatic differentiations, integrate functions, and even solve differential equations. By embedding such information, the amount of data required for training is reduced. These techniques will be essential for the later treatment of

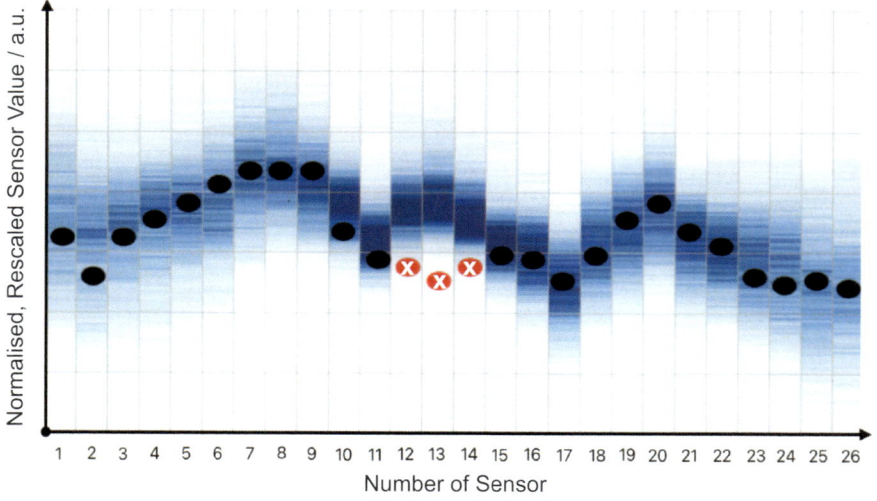

Fig. 6.14 Process corridor for several decoupled sensors, correspondingly normalized and strung together. If an element runs through the unlikely area multiple times, an anomaly is present. Black circles show the respective variable. Red circles with white crosses show the anomalous variables. They deviate too far from the expected distribution, which is illustrated shaded in the background here

explainable machine learning—which is based on a thorough understanding of data enrichment and the inclusion of prior knowledge.

Uncertainty can be captured with the help of mixture-density networks and systematically integrated into the prediction of model results. Other methods such as decision trees or clustering methods can also be expanded to include the concept of uncertainty. Stochastic enrichment of data allows, with knowledge of the underlying uncertainties, to include these in a practical way in the evaluation.

Particularly interesting for technical processes is the concept of the process corridor. Here it was shown how different distributions can help to monitor processes for anomalies.

Tasks

6.1 Consider again our example of the motor current. In Sect. 3.6.3 we got to know the FFT and in Sect. 6.1.2 we explained how adding a Fourier transformation affects the classification quality of the neural network. Implement a physically-informed classification for the motor current example with a neural network. Use the following preprocessing:

a) Fourier Transformation as preprocessing

b) Laplace Transformation, which is defined over

$$\mathcal{L}(\zeta) = \int_0^\infty x(t)e^{-\zeta t}dt$$

.

What difference do you recognize with regard to the classification?

6.2 Sketch the database concept so that a physically-informed code for machine learning can automatically retrieve enrichment information for the prediction of water pumps. Which variables are important?

6.3 Program the integration of the function

$$f(t) = 4t + \cos(2\pi t)$$

using a neural network.

6.4 Write the ordinary second order differential equation

$$x'' = 2x + t$$

as a system of first order differential equations. Adjust the code from Listing 6.8 and 6.11 so that second order differential equations can also be integrated with the first order code.

6.5 How would you apply the principles of physically-informed machine learning to the K-Means method that you learned in the previous chapters?

References

1. C. M. Bishop, *Neural Networks for Pattern Recognition*. Oxford University Press, 1996.
2. C. M. Bishop and D. Barber, „Ensemble learning for multi-layer networks," in *Advances in Neural Information Processing Systems*, vol. 10, 1997, pp. 395–401.
3. A. Brando, „Mixture density networks (mdn) for distribution and uncertainty estimation," 2017, gitHub repository with a collection of Jupyter notebooks intended to solve a lot of problems related to MDN. [Online]. Available: https://github.com/axelbrando/Mixture-Density-Networks-for-distribution-and-uncertainty-estimation/.
4. D.Barber and C. M. Bishop, „Ensemble learning in bayesian neural networks," in *Generalization in Neural Networks and Machine Learning*. Springer Verlag, 1998, pp. 215–237.
5. S. J. Julier, T. Bailey, and J. K. Uhlmann, „Using exponential mixture models for suboptimal distributed data fusion," *IEEE Nonlinear Statistical Signal Processing Workshop*, pp. 160–163, 2006.
6. G. E. Karniadakis, I. G. Kevrekidis, L. Lu, P. Perdikaris, S. Wang, and L. Yang, „Physics-informed machine learning," *Nature Reviews Physics*, vol. 3, pp. 422–440, 2021.
7. I. E. Lagaris, A. Likas, and D. I. Fotiadis, „Artificial neural networks for solving ordinary partial differential equations," *IEEE Transactions on Neural Networks*, vol. 9, no. 5, pp. 987–1000, 1998.
8. J. Maggu, A. Majumdar, E. Chouzenoux, and G. Chierchia, „Deep convolutional transform learning," in *ICONIP 2020—27th International Conference on Neural Information Processing, Bangkok, Thailand*, 2020.
9. R. G. Nascimento, K. Fricke, and F. A. Viana, „A tutorial on solving ordinary differential equations using python and hybrid physics-informed neural network," *Engineering Applications of Artificial Intelligence*, vol. 96, p. 103996, 2020. [Online]. Available: https://www.sciencedirect.com/science/article/pii/S095219762030292X.
10. M. J. Neuer, *Quantifying Uncertainty in Physics-Informed Variational Autoencoders for Anomaly Detection*. Springer Nature, 2021.
11. D. Pfau, J. S. Spencer, A. G. Matthews, and W. M. C. Foulkes, „Ab initio solution of the many-electron schrödinger equation with deep neural networks," *Phys. Rev. Res. 2*, vol. 2, p. 033429, 2020.
12. M. Raissi, P. Perdikaris, and G. E. Karniadakis, „Physics-informed neural networks: a deep learning framework for solving forward and inverse problems involving nonlinear partial differential equations," *J. Comput. Phys.*, vol. 378, pp. 686–707, 2019.

Chapter 7
Explainability

Keywords Semantic structures · Taxonomy · Ontology · Vocabulary · Synonyms · Explainability · Causality

To make learning processes comprehensible, we need to model input variables, process characteristics, and relationships between the variables. For this purpose, taxonomies and ontologies are introduced. They are tools to create structured information storage for physics-informed learning and explainability. The chapter is dedicated to sensitivity analysis and explains with an example how learned process models can be examined for their dependencies. Finally, a discussion of explainability and a clear recipe for how to implement it in your own projects concludes this chapter.

The semantic tools of this chapter enable us to generate exactly this explainability for our learning processes, depending on user types and groups of people. However, a new form of information must first be digitally captured: the "meaning" of procedures and their associated process know-how. This meaning is ultimately needed for the interpretation of results.

Fig. 7.1 again shows the effects of Chap. 7, with a large part now addressing the user of our procedures.

7.1 Semantic Ordering Structures for the Digitization of Meaning

We have already encountered many different examples of dealing with data. In all these cases, it was about analyzing, evaluating, classifying existing measurement data, and ultimately making a statement about these data.

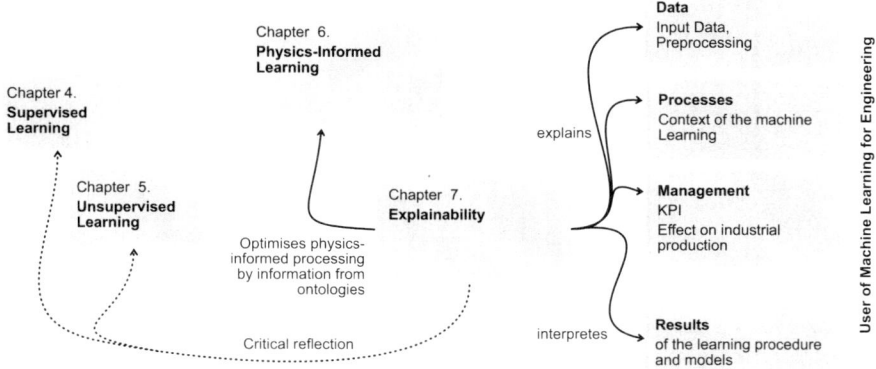

Fig. 7.1 Overview of the relationship of Chap. 7 with the previous chapters

In this chapter, we will focus on a meta-modeling context. Why is this important? Computers initially have no access to information about relationships. They certainly do once mature AI algorithms interact with the World Wide Web, but for the moment we want to remain in the field of technical applications, e.g., for industry.

Does a computer know what pickling is? Does it know the meaning of an ultrasonic test? It knows the data of these processes, but it cannot place them in larger overall contexts. This step has, at least in all our examples, only been taken by the human data scientist. The specialist decides the preprocessing of the data, selects the procedure, and interprets the results—not the computer.

7.1.1   Semantics

Semantics describe the meaning of things. When we talk about a "car", everyone we talk to knows what we mean: a vehicle that can move from A to B, consumes energy, can hold one to five passengers, etc. You know the meaning of the word "car" because you have learned what it is.

A computer could do the same, but it has never gone through this learning process. For example, you can train an image recognition system that can distinguish a "penguin" from an "elephant"—but the interpreter will not understand what a penguin is just by this distinction.

Now imagine there is a table in which a "penguin" is assigned to the superordinate category "bird". Then an algorithm would understand that a penguin is a bird. But now the term "bird" needs to be explained.

A completely digital capture of real objects therefore requires us to find a way to provide computers with digital explanations of these objects. And this is where the semantic tools **synonym, taxonomy** and **ontology** come into play. All of these are tools and technologies to digitally store the **meaning** of something and make this information usable for algorithms.

7.1.2   Overview of Semantic Concepts

In Fig. 7.2 we have listed the most important concepts of semantic ordering structures and arranged them in order of increasing complexity. We will systematically explain these concepts in the following section. They are important for the explainability of algorithms. Our overarching goal is to build explainable machine learning in a technical environment. How can these concepts help us with this?

> **Example: Anomaly Detection with Semantic Assignment**
> A hydraulic press compresses strips in a process. This results in embossed good parts for further processing. The force (or pressure) of the press is continuously and digitally stored; it is also visualized via an online display. Based on the curve, an expert can determine whether the process was in order (i. o.) or not in order (n. i. o.). Now, whenever the maximum press force is expected, a sudden drop in the system's force is detected. This anomaly can (for our example!) be traced back to two possible causes: 1) Either the basic performance of the hydraulic pump briefly drops or 2) a crack in the supply line causes fluid to leak. For data purposes, the hydraulic pump is also digitally recorded.
> The semantic solution in this case is possible as follows: We know that the pump is necessary for generating the force. Pump and press are therefore logically connected. We can semantically describe this with the following logic: If the pump pressure drops, so does the press force. This if-then relationship is the core element of our consideration. So if the pump signal shows that there was no performance breakdown, then only cause 2), a leak in the supply line, can be the possible explanation. ◄

This example is based on so-called **deontic logic.**

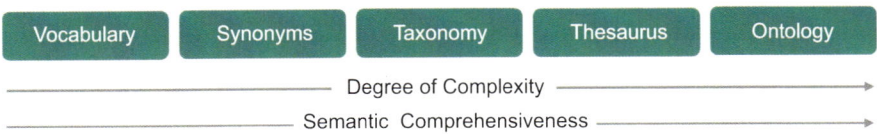

Fig. 7.2 Semantic tools, ordered according to increasing complexity

7.1.3 Alphabets and Words

How do we form words at all and when does a word contain meaning? A word consists of many small units, the letters. The special thing about these symbols is that they all have to be easily distinguishable from each other. Otherwise, there would often be confusion.

> **Alphabet** An alphabet Σ is a finite set of clearly distinguishable elements $s_i \in \Sigma$. Such an element s from Σ is called a character (symbol, letter).

A word can now be constructed from the elements of an alphabet:

> **Word** Every finite combination of characters of an alphabet $x, y, z \in \Sigma$ is called a word. The number of characters determines the length of the word.

This view of computer science does not yet include any sense that a word can carry. All possible combinations of characters are words. Ultimately, the assignment of meaning is the main task of the term semantics:

> **Semantics** If a meaning is assigned to a word, this is referred to as a semantic mapping.

Only through this process are special words distinguished from the set of all words that can be formed by the alphabet Σ: they are given a meaning. Through this step, we obtain a subset of the possible words of the alphabet Σ. However, two different words can also contain the same meaning. We will revisit this property later in our consideration of synonyms.

7.1.4 Vocabulary

If you have ever learned a foreign language, you are certainly generally familiar with the meaning of vocabulary. It describes a collection of words regarding a common context and in a common language. In computer terms, a vocabulary can be an array of strings. With the help of the vocabulary, it can then be checked whether an entered text is actually compatible with the respective language.

> **Vocabulary** A vocabulary is the set of words that can be formed from an alphabet Σ to which a meaning has been assigned through a semantic mapping. A **limited vocabulary** limits these words to a certain context of meaning.

7.1.5   Synonym Ring

Do you say "I ride a bicycle." or "I am biking."? Apparently, different words can have the same or at least a very similar meaning.

> **Synonym** Two words whose semantic assignment of meaning is the same are called synonyms.

In computer science, the handling of synonyms is well known. When string-based inputs are evaluated, a synonym table is helpful. Let's discuss this again using a practical example:

> **Example: Defects in Steel Strips**
> Steel strips can have various defects. One of these defects relates to the edge of the steel strip. Here, cracks can occur due to the process. This is problematic for subsequent processes and often the starting point for a complete strip tear—which should be avoided as much as possible. If you now observe strips in production, you can of course assign an employee to manually inspect the strip edge and write down whether there is a problem or not. What you will get as soon as a crack is seen is a table with data entries like "edge tear", "edge crack", "fraying", "notch" and much more. For the underlying problem, these are all synonyms for the actual defect. ◄

When a variety of terms exist that describe a fact with the same meaning, we call these terms synonyms. We say, one term is synonymous with another term of the set. Mathematics knows this as the concept of relation, more specifically the equivalence relation. It allows us to establish the reference of elements in a set.

Equivalence relation Let there be elements $A \in S$ and $B \in S$. We call $\equiv$ an equivalence relation on S, if every element $x_i \in S$ is equivalent to itself $x_i \equiv x_i \forall i$ (reflexivity), if from $A \equiv B$, $B \equiv A$ follows (symmetry) and if from $A \equiv B$ and $A \equiv C$ follows, that $B \equiv C$ is (transitivity).

An equivalence relation must therefore be reflexive, symmetric, and transitive.

Are $A \in S$ and $B \in S$ words of a word set S, on which an equivalence relation $\equiv$ is defined, then all $x_i, x_j \in S$ with $i \neq j$, which are equivalent to each other, $x_i \equiv x_j$ form a synonymous ring Ξ.

Example: Synonymous Ring for the Word Automobile
The synonymous space for the German word "automobile" is given by

$$\Xi(\text{„Automobil"}) = \left\{\text{„PKW"}, \text{„Auto"}, \text{„Kraftfahrzeug"}, \text{„Wagen"}, \ldots\right\}. \quad (7.1) \blacktriangleleft$$

The example in Eq. (7.1) can be implemented in Python. For this, we have programmed this word set and an example text in Listing 7.1. From the example text, the loop should find out how often the term "automobile" or one of its synonyms occurs.

Listing 7.1 Synonym analysis

```
Synonyms = {"Automobil":["Auto", "PKW", "Kraftfahrzeug", "
    Wagen"]}
Counter = {}

for eachWord in Text.split('_'):
    for eachEntry in Synonyms:
        if eachWord in Synonyms[eachEntry]:
            if eachEntry not in Counter.keys():
                Counter[eachEntry] = 1
            else:
                Counter[eachEntry] +=1

print(Counter)
```

Please remember our consideration of data types and the characterization of data. A text represents a collection of nominal data. And for nominal data, we can only count the occurrences of an element, nothing more. That's why the above example is equipped with a counter. The number of occurrences of the synonyms for "automobile" is the only operation we can perform.

7.1.6 *Taxonomies*

The term taxonomy (Greek *táxis,* the order) describes a rule with which objects are grouped. In doing so, one looks for an element that connects all objects and defines this as the superordinate term of the taxonomy.

> **Example: Simplified Biological Taxonomy**
> In biology, taxonomies have been known for a long time. Fig. 7.3 shows a simplified order structure for a part of the animals. Only the two categories vertebrates and arthropods are considered. Vertebrates are further divided into mammals, birds, fish, and amphibians. A dog belongs to the category mammal. However, further detail levels of the dog are distinguished. ◄

If the information from Fig. 7.3 is digitally available, an automatic procedure can algorithmically conclude that a dog is an animal. Without the dependency tree of the taxonomy, this is not possible. Order structures require a criterion, i.e., a mathematical rule, according to which their elements can be sorted. This sorting brings the required order.

In comparison with the synonymous ring, which merely requires an equivalence between its elements, a taxonomy can therefore compare. Taxonomies convert nominal data into ordinal data.

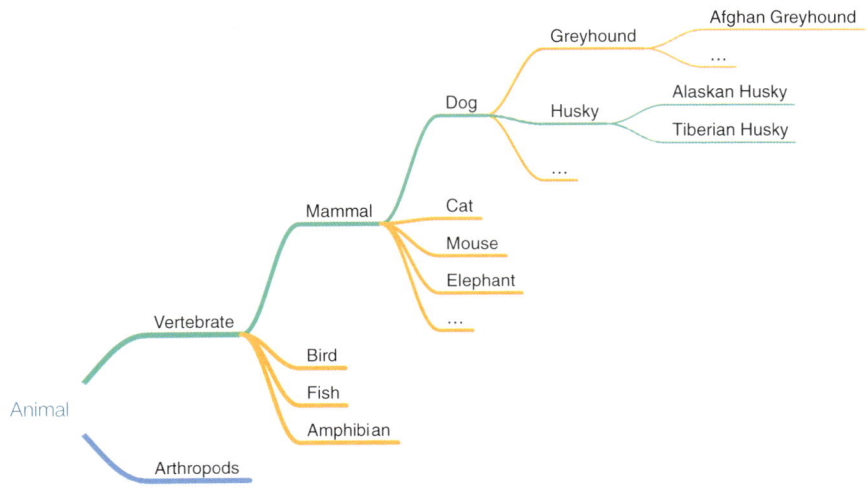

Fig. 7.3 Simple example of a taxonomy, based on biology

Order relation An ordered set M is a set for whose elements $x, y, z \in M$ applies:

$$x \leq x \text{ Reflexivity,} \tag{7.2}$$

$$x \leq y \wedge y \leq x \Leftrightarrow x = y \text{ Antisymmetry,} \tag{7.3}$$

$$x < y \wedge y < z \Leftrightarrow x < z \text{ Transmittivity.} \tag{7.4}$$

On this ordered set, $<, >, \leq$ and $\geq$ are order relations for the elements. This leads us to a compact and clear definition of a taxonomy:

Taxonomy A taxonomy is an ordered set of terms. The order relation on these terms arises from the context given to them.

Let's illustrate this using the example of the term "automobile", which we have just used to explain a synonymous ring to us. But now we expand the terms so that different vehicle sizes are covered.

Example: Taxonomy for Automobiles
The taxonomy of road users sorts all its elements according to their mass. Then the ranking within the taxonomy is: 1) Trucks larger than 20t, 2) Trucks smaller than 20t but larger than 7.5t, 3) Small trucks up to 7.5t, 4) Cars, 5) Motorcycles, 6) Bicycles and 7) People. ◄

To understand how a programmed taxonomy works, we have set a naive hierarchy in Python in Listing 7.2. It shows you a way to use synonym sets to compare different terms with each other.

Listing 7.2 Taxonomy

```
import numpy as np
class taxonomy(object):

    def __init__(self, name):
        self.name = name
        self.set = {}
        self.entries = []
        self.index = []

    def addSynonymList(self, dictOfSynonyms, orderNumber):
        for eachEntry in dictOfSynonyms:
            self.set[eachEntry]=dictOfSynonyms[eachEntry]
            self.entries.append((str(eachEntry),orderNumber))

    def hierachy(self):
        dtype = [('name', 'S10'), ('index', int)]
        data = np.array(self.entries, dtype=dtype)
        return np.sort(data,order='index')

    def isLarger(self, element, comparison):
        for eachElement in self.set:
            if element in self.set[eachElement]['Synonyms'] or
                element in eachElement:
                    theElement = eachElement
            if comparison in self.set[eachElement]['Synonyms']
                or comparison in eachElement:
                    theComparison = eachElement

        for eachTuple in self.entries:
            if theComparison in eachTuple[0]:
                theIndexOfComparison = eachTuple[1]
            if theElement in eachTuple[0]:
                theIndexOfElement = eachTuple[1]

        if theIndexOfElement>theIndexOfComparison:
            return True
        else:
            return False

myTaxonomy = taxonomy(name='Verkehrsteilnehmer')
myTaxonomy.addSynonymList({"Automobil":{"Synonyms":["Auto", "
    PKW", "Kraftfahrzeug", "Wagen"]}},2)
myTaxonomy.addSynonymList({"Fahrrad":{"Synonyms":["Bike", "Rad
    ", "Drahtesel", "Zweirad"]}},1)
myTaxonomy.addSynonymList({"LKW":{"Synonyms":["Laster", "
    Lastwagen", "Truck"]}},3)

print(myTaxonomy.isLarger('Laster','Rad'))
print(myTaxonomy.isLarger('Drahtesel','LKW'))
```

As a result, this taxonomy doesn't care which term from the synonym list you provide for comparison. In line 43 we compare "truck" and "wheel". Since the truck actually has the order number 3, it is larger than a bicycle in this order structure. In practice, such word sets are stored in large databases. Fast queries then allow similar processes as shown in the code.

7.1.7   Ontologies

Modeling of Relationships

Ontologies contain concepts, similar to taxonomies. However, they are also capable of capturing relationships between these concepts. The simultaneous modeling of a semantic description of an object and its relationships to other objects characterizes ontology. It can digitally store knowledge and context.

> An **ontology** $O(K; I, R)$ is a space with the classes K, objects I that are instantiated from these classes, and relations R that represent relationships between the objects.

Only when a computer system has an ontology can it evaluate complex relationships from information. In the article by Babik et al. from 2006 [1] an integration of the Web Ontology Language (OWL), see also [8], in Python is shown. How important the role of ontologies can be is made clear in the works of Furbach et al. [4, 5], where the authors show ways in which automatic decision-making can benefit from an ontology.

We explain this using an example:

> **Example: Continuous Casting**
> Steel is cast into a slab in continuous casting. It consists of a ladle with liquid steel, a distributor, a mold, the guide rollers, a plug, and many other components. Various variables influence this process: the chemical composition of the steel, the casting speed, the temperature of the steel melt, the temporal behavior of the bath level, a possible blockage of the casting outlet, and various actuators, such as the amount of argon added.
> All these influences are thus related to the continuous casting process. These variables are responsible for its success. ◄

A useful tool for creating and managing ontologies is Protegé [7]. It offers a graphical interface that significantly simplifies handling semantic properties and creating relations. Import and export of various ontology formats are also integrated here. The Semantic-Media-Wiki[1]is also a form of ontological data management.

Implementation of a micro-ontology in Python

---

[1] https://www.semantic-mediawiki.org/

To get to know the basic concept in a more practical way, we construct our own small ontology, a micro-ontology. We use the continuous casting example as a basis. In Listing 7.3 the structure of a class called `ProcessElement` is shown.

Listing 7.3 Micro-ontology

```
class ProcessElement():

    def __init__(self, inName:str='', inId:int=0):
        self.name = inName
        self.iAmPartOf = []
        self.iContain = []
        self.iCanProduce = []
        self.id = inId

    def addPart(self,x):
        self.iContain.append(x)
        x.beingPartOf(self)

    def beingPartOf(self, y):
        self.iAmPartOf.append(y)

    def showWhatIAmPartOf(self):
        print(['{}_({})'.format(element.name, element.id) for
            element in self.iAmPartOf])

    def showWhatIContain(self):
        print(['{}_({})'.format(element.name, element.id) for
            element in self.iContain])

Caster = ProcessElement(inName='Strangguss')
for i in range(0,10):
    newGuide = ProcessElement(inName='Fuehrungsrolle', inId=i)
    Caster.addPart(newGuide)

Mold = ProcessElement(inName='Kokille')
Caster.addPart(Mold)

Caster.showWhatIContain()
Mold.showWhatIAmPartOf()
```

We then instantiate this class to capture the various objects in the process and connect them. Using the terms `beingPartOf()` and `addPart()` we can model a process. In the code, we have created a `Caster` that contains 10 guide rollers as parts and a mold.

Micro-ontology for physically-informed learning
In Listing 7.4 we show how you can extend the ontological description. For this, we introduce our own description level for variables, which we call `Quantity`. It captures not only the name of the variables, but also the information about uncertainty, distribution, unit, and their preprocessing steps (`modes`).

Listing 7.4 Micro-ontology for input variables

```
class Quantity():

    def __init__(self,
                  inName:str='',
                  inUnit:str='',
                  inDistribution:str='',
                  inUncertainty:float='',
                  inModes=[]):
        self.name = inName
        self.unit = inUnit
        self.distribution = inDistribution
        self.uncertainty = inUncertainty
        self.modes = inModes

    def getModes(self):
        return self.modes


class ProcessElement():

    def __init__(self, inName:str='', inId:int=0):
        self.name = inName
        self.iAmPartOf = []
        self.iContain = []
        self.iCanProduce = []
        self.myQuantities = []
        self.id = inId

    def addQuantity(self, q):
        self.myQuantities.append(q)

    def addPart(self,x):
        self.iContain.append(x)
        x.beingPartOf(self)

    def beingPartOf(self, y):
        self.iAmPartOf.append(y)

    def showWhatIAmPartOf(self):
        print(['{}_({})'.format(element.name, element.id) for
            element in self.iAmPartOf])

    def showWhatIContain(self):
        print(['{}_{})'.format(element.name, element.id) for
            element in self.iContain])

    def showMyQuantities(self):
        for eachElement in self.iContain:
            for eachQuantity in eachElement.myQuantities:
                print('{}({})'.format(eachQuantity.name,
                    eachQuantity.getModes()))
```

```
50  Caster = ProcessElement(inName='Strangguss')
51  for i in range(0,10):
52      newGuide = ProcessElement(inName='Fuehrungsrolle', inId=i)
53      Caster.addPart(newGuide)
54
55  Mold = ProcessElement(inName='Kokille')
56  MoldLevel = Quantity(inName='Giesspiegel', inUnit='mm',
        inUncertainty=0.2, inDistribution='Gauss', inModes=['FFT',
        'Derivative'])
57  ArgonFlow = Quantity(inName='ArgonFluss', inModes=['Maximum',
        'Minimum','Log'])
58  Mold.addQuantity(MoldLevel)
59  Mold.addQuantity(ArgonFlow)
60
61  Caster.addPart(Mold)
62  Caster.showWhatIContain()
63  Mold.showWhatIAmPartOf()
64
65  Caster.showMyQuantities()
```

This short example is intended to show you how you can quickly build local ontologies. However, it is always desirable to carefully and structurally create a complete ontology for an entire process environment. The above approaches would then only be part of such a larger description of the system.

7.2   Sensitivity Analysis

To make models understandable, we need to look inside a black-box model. Our goal in this section is to arrive at a statement about their sensitivity to the model prediction by successively sampling the input variables. Bastani et al. show such an approach with a focus on a decision tree in [2]. We will choose a similar path, but keep it simple and justify it through a perturbation theory.

7.2.1   Perturbation Theoretical Approach

In some disciplines of physics, perturbation theory refers to deliberately introducing a disturbance into a model and then determining the system's response. This approach has some strong limitations. The first limitation is that it is a small disturbance. This is expressed by a so-called smallness parameter ε, for which $\varepsilon \ll 1$ applies. Suppose we have a test vector x and we disturb it, we write

$$\xi = x + \varepsilon \delta x + \varepsilon^2 \delta x_2 + \mathcal{O}(\varepsilon^3) \tag{7.5}$$

and obtain a disturbed state vector $\boldsymbol{\xi}$. We only carry out the quadratic disturbance component for illustration and will focus on the effect of $ve\delta\boldsymbol{x}$. Now let $\mathcal{F}$ be any learning method, we estimate the influence of the disturbance as follows,

$$
\begin{aligned}
\boldsymbol{y} &= \mathcal{F}(\boldsymbol{\xi}) \\
&\approx \mathcal{F}(\boldsymbol{\xi}) + \varepsilon\mathcal{F}(\boldsymbol{\delta x}) + \varepsilon^2\mathcal{F}(\boldsymbol{\delta x_2}) + \mathcal{O}(\varepsilon^3),
\end{aligned}
\tag{7.6}
$$

where another strong assumption becomes visible, namely that a linear, additive disturbance in $\boldsymbol{\delta x}$ has an additive effect on $\boldsymbol{y}$. Bhatt et al. also use a selected, central test point in [3] and extend the concept to a generalized explainability. Current works, such as those by Tan et al. [11], deal with the question of how much the restriction on additivity affects such approaches.

We limit ourselves in the version discussed here to the simplified approach above. The reason for this is that it leads to good results in many practical applications and allows us to convey the ideas behind the word explainability mathematically. The concept can also be directly applied to all the learning methods presented here.

For further ideas and methods, we would like to first point out the LIME and SP-LIME approaches by Ribeiro et al. in [9], which also consider the redundancy of input variables. Furthermore, there are several works, e.g. [6] or [12], which go back to the so-called Shapley values and are based on a foundation of game theory by L. Shapley [10]. There are separate Python packages for both analysis strategies, LIME and SHAP, which can easily be applied to the examples presented in this book, but a discussion of which would go too far at this point.

7.2.2 Perturbation of a Decision Tree

To fill the above theoretical formulation with life, let's consider a simple but instructive example. For this, we first generate a set of synthetic data. These are chosen in such a way that we can easily apply the disturbance and manually check whether our disturbance delivers the correct statements based on the chosen type of data.

We generate the synthetic data with a function

$$
y = \mathcal{F}(\boldsymbol{x}) = x_0 + x_1^2 + \log(1 + x_2),
\tag{7.7}
$$

that maps an input vector x with three components $x = (x_0, x_1, x_2)$ to a scalar value y. As you can see in (7.7), this mapping is a nonlinear combination of the components. Listing 7.5 shows the code for generating the training data.

Listing 7.5 Synthetic data for demonstrating perturbation approach

```
import matplotlib.pyplot as plt
import numpy as np

def function(x):
    return x[0]+x[1]**2+np.log(1+x[2])

def generateSyntheticData(n):
    X = []
    Y = []
    for i in range(0,n):
        x = 5*np.random.random(3)
        y = function(x)
        X.append(x)
        Y.append(y)
    return X, Y

Xtrain, Ytrain = generateSyntheticData(10000)
```

It is important to note that in line 11 a random x vector is generated and then Eq. (7.5) is applied to it. For x we use values between 0 and 5.

We now train a decision tree from Scikit-Learn to establish a model from the training data. Listing 7.6 shows the call of this training.

Listing 7.6 Fit using decision tree

```
from sklearn.tree import DecisionTreeRegressor

reg = DecisionTreeRegressor()
reg.fit(Xtrain, Ytrain)
```

With the synthetic data and our model, we have laid out all the pieces for understanding a practical sensitivity analysis. In Listing 7.7 we create a function `perturbationScan`. It is passed the decision tree `reg` and the disturbance position `pos`. We do not vary x_0, but only the shares x_1 and x_2 of x. The variation δx is added to a ground state, we choose this ground state as $(0, 0, 0)$.

Listing 7.7 Perturbation scan for sensitivity analysis

```
def perturbationScan(reg, pos=0, scanRange=np.arange
    (0.1,5,0.2)):
    y = []
    x = []
    yt = []
    for i in scanRange:
        x0 = np.array([0.0,0.0,0.0])
        deltaX = np.array([0.0,0.0,0.0])
        deltaX[pos] = float(i)
        test = [x0 + deltaX] # x0 + eps*x1 ...
        yr = reg.predict(np.array(test))
        x.append(i)
        y.append(yr[0])
        ytest = function(test[0])
        yt.append(ytest)
    return x, y, yt

x1, yfx1, testfx1 = perturbationScan(reg, 1)
x2, yfx2, testfx2 = perturbationScan(reg, 2)
```

In Fig. 7.4, the result of this analysis is shown. The reference `testfx1`and
`testfx2` are plotted as a black line. The red crosses symbolize our scan result
`yfx1` and `yfx2`. So, we actually obtain the true functional relationship that we

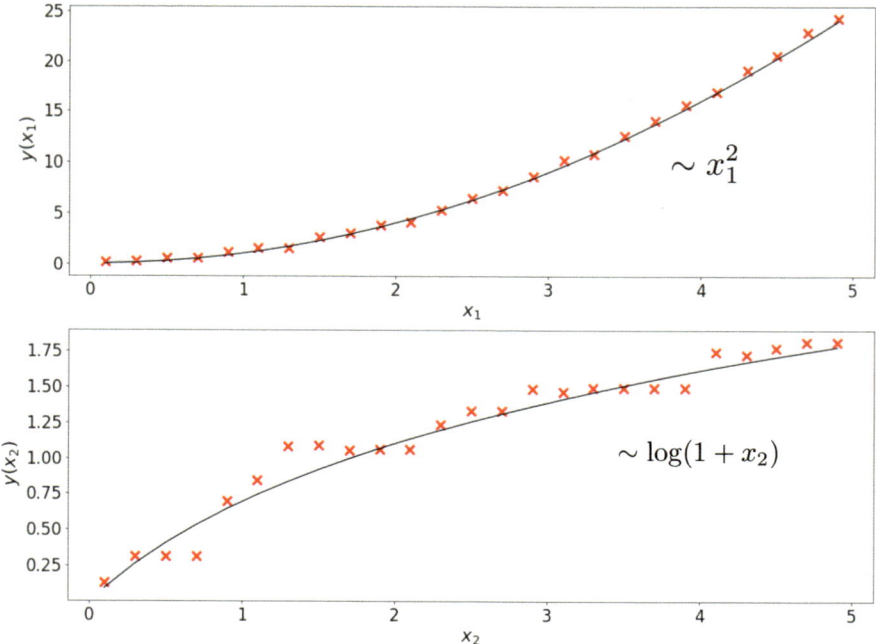

Fig. 7.4 Result of the sensitivity analysis from Listing 7.7

originally defined for our data generation. In this way, you can systematically investigate learning methods and analyze their dependence on the input variables.

7.3 Explainable Machine Learning

Explainability is becoming increasingly important for algorithms. In the various chapters, we have therefore laid out the components to realize explainability. We now bring these individual components together to form a picture for explainable learning in the technical environment.

7.3.1 Levels of Explainability

When we explain something to someone, then it is crucial to know the listener's prior knowledge and the context of the explanation well. The same applies to algorithms. Depending on the group of people, the explanation must be tailored appropriately. We now have the following levels of explainability that we can distinguish, and each of these levels can (but does not have to) correspond with different people and questions:

(E1) **Business level.** At the beginning of every data-based project is the analysis of its meaningfulness. Can a model help the process become faster, more qualitative, or more robust, and if so, what monetary and administrative advantage does this have? The user type here are managers and decision-makers.

(E2) **Process level.** At this level, we consider the technical relationships of the actual problem. Is it a paper press, an engine, or a complex combination of several processes? The relevant group of people here includes process experts who have deep knowledge of the actual process and plant personnel who work daily, for example, with a manufacturing process.

(E3) **Data level.** In preprocessing, data is altered with the aim of making it more interpretable and clearer. The proverbial "glasses" with which an algorithm sees the results better also requires special know-how. The relevant group of people here is the data scientist, but also process experts who provide feedback on data processing.

(E4) **Model level.** Which algorithm was selected and why? Can we derive further statements from the model? The primary user at this level are data scientists.

(E5) **Result level.** How meaningful, substantial, and secure are the results? Was the uncertainty taken into account? Within what framework does the model's prediction apply? The group of people are decision-makers and process experts, so the explainability of the results is perhaps the most important.

(E6) **Causal level.** Is the model causally meaningful? This affects all groups of people and can often only be evaluated by groups.

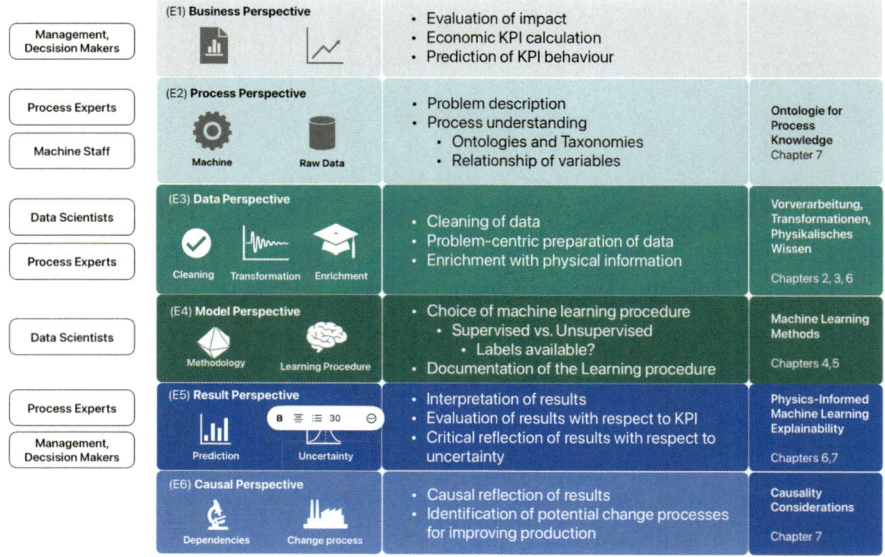

Fig. 7.5 Overview of the different levels of explainability

Fig. 7.5 summarizes the levels visually. It also connects the various topics of the previous chapters with these levels, which we will go into in more detail below.

7.3.2   Practical Implementation of Explainability

Ontology of the Process Level

The process level is often represented by a group of people with deep process know-how. Answers from machine learning methods should tie in with the technical experience of these experts and be especially comprehensible. If a model delivers results, process experts must be sure that the predictions are consistent. The most important element of the process level is the raw data. It must be ensured that these provide an unadulterated picture of the process and that the process experts trust these data.

Please note that you have no influence on the trust of such an expert in his raw data from the modeling side. If this trust does not exist, no trust can be placed in a model based on these data. So make sure that the raw data is reliable and that this is a consensus of all participants.

First, construct an ontology with the help of process experts. A multi-step approach is advisable here. Start with the most important processes and find descriptions for their mechanisms of action. Abstract and reduce the content as much as possible. Here, relationships and existing laws can help.

> **Example: Reduced, abstracted description of the rolling process**
> Goal: Deformation; Physical actors: Force, Pressure, Temperature. ◄

Once you have captured the individual processes, model the relationships between the processes. Fig. 7.6illustrates an example of this. Now the actual ontology emerges from this. You now have a digital process description that can be queried by your other components.

Ontological enrichment of physical interactions
Based on this initial ontological description, you now supplement the data objects in your ontology with the description of physical interactions. Every data variable that is relevant for your analysis project should have a digital representation here. Not every variable necessarily has to be a physical quantity. You can of course structurally expand new quantities as required by your individual process environment.

This is shown on the right side of Fig. 7.6. You have now reached an important intermediate stage for explainability: the process level and data level are digitally explainable.

Explainability of preprocessing
Every preprocessing step that increases the understanding of data is worth explaining. But how can a learning method automatically extract from an ontology the fact that the CWT could be a good preprocessing for a certain process variable? The answer is by considering the physical mechanism of action. If we explicitly

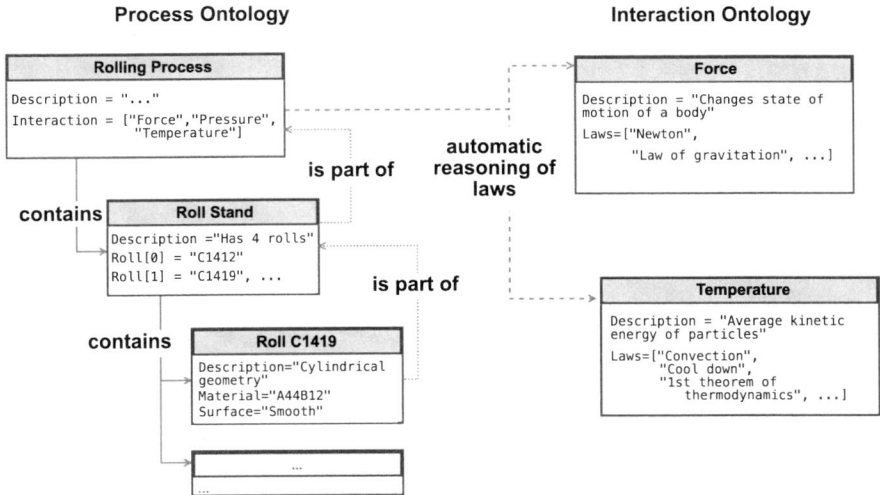

Fig. 7.6 Combination of process ontology and description of interactions

add the information about their best transformations to a variable, we can, for example, directly infer the preprocessing "derivation" for the variable type "rolling force". The comprehensibility of preprocessing lives from the justification of its individual steps. We find this justification in the description of the variables.

The extraction of characteristic features, see Sect. 3.8, can also be stored in an ontology in this way. For this, an additional entry of the dependency in the process is added for each feature and the feature itself is inserted as a dependent variable in the ontology. It would then stand on the right side in Fig. 7.6 and be included on the left side as a dependency in the process description.

Automation of physics-informed learning
The procedure for an automatic analysis chain is as follows: The input variables have a unique name. It refers to a quantity in the ontology. This either has a physical context or not, but in any case it is described there. You now call up the data element of the ontology that describes the variable and then have access to a description for your data and a reference to suitable preprocessing steps.

Then, You select the transformations that were in the ontology for your variable. These are passed to the learning method, along with the data itself. Human knowledge has thus been digitized via the ontology and can be used by the learning method. This step increases the intelligence of the entire system, as the learning method no longer has to train the best processing itself.

The procedure can of course also be applied equally to the integration of differential equations into the training process and to the impact of uncertainties. For this, we again look at the right side of Fig. 7.6. Here, the description of the force also contains a reference to the law of gravity. An analysis process can now automatically query this relationship and take it into account in its evaluation.

Deductive element for explaining the result
There is now another aspect that completes our consideration of explainability: deductive reasoning from semantic information. Let's look at Fig. 7.7, which illustrates a larger context of the entire process for the following example.

Example: Cause-Effect Deduction for an Oscillating Circuit
A circuit consisting of a capacitor, coil, and resistor acts as an oscillating circuit. We now measure both the current and a quality parameter—e.g., the noise—which tells us how good our circuit is. With a neural network, we are already able to deduce problems in quality from the temporal current flow.
We now use a physically informed ontology. It contains the meaning of the current strength and its relationships to other quantities, such as resistance or voltage. Two transformations are proposed for the variable type "current": FFT and derivative (Derivative). Both are now used as preprocessing steps to enrich the data of the learning process.
With a subsequent sensitivity analysis, we finally find out that the FFT inputs of the neural network have the strongest influence on the result. ◄

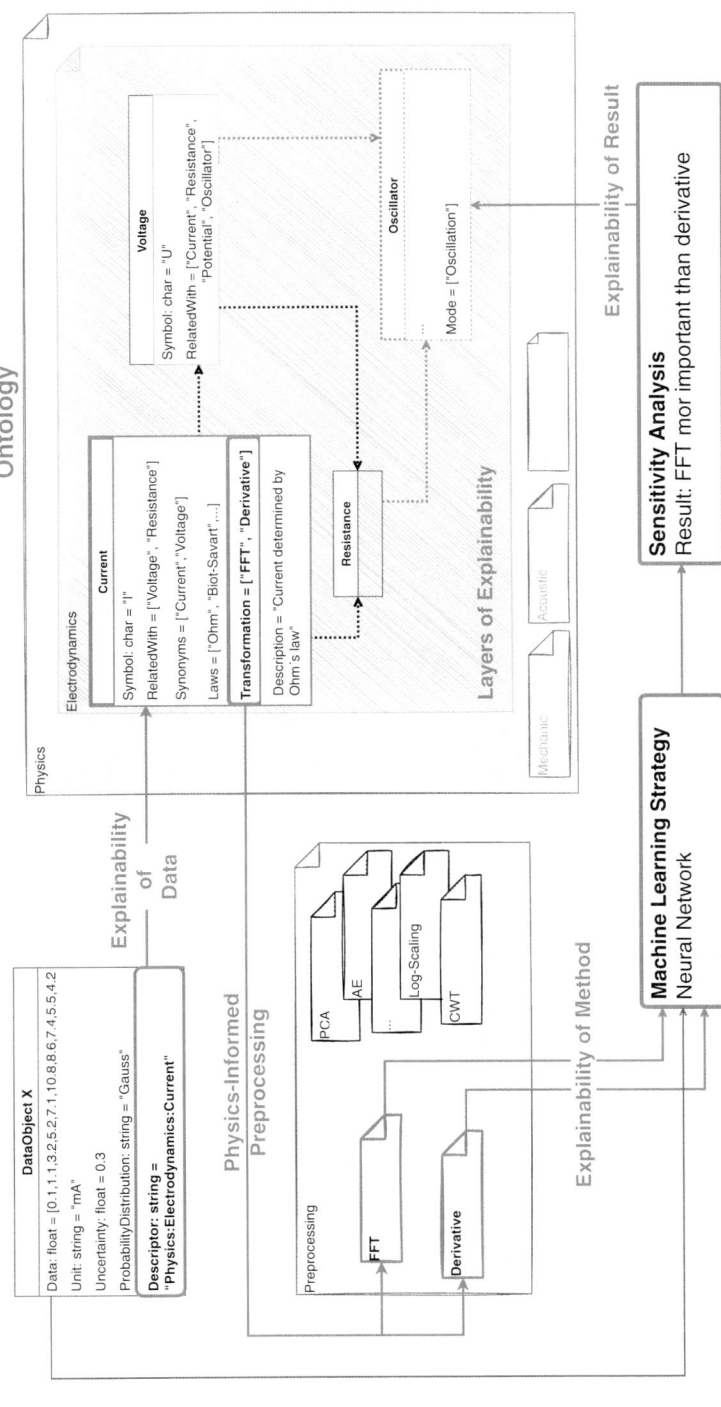

Fig. 7.7 Connection of data, ontology, physics-informed preprocessing, learning method and sensitivity analysis, with the aim of achieving explainability at all levels

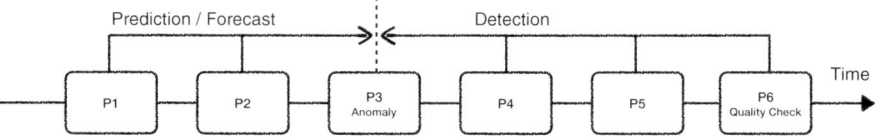

Fig. 7.8 Causal flow through an abstract process chain. P1–P5 are processes that interact with a product and P6 is a specific quality check

If a derivative had the strongest influence, we would have to assume a shock-like disturbance. However, since the FFT is linked to the oscillations in the system, the error considered here can only originate from the components of the oscillating circuit. At this point, a deductive conclusion is made. From the general semantic knowledge in the ontology, it can be directly concluded in the specific example which assembly expresses the error.

Even though the example shown here is greatly simplified, it hopefully gives you an impression of the logical conclusions that semantic digitization of information can help us achieve.

7.3.3 Causality in the Technical Process Chain

In manytechnical situations, we are indeed interested in understanding the relationship between cause and effect, i.e., uncovering the causal connection.

> **Example: Surface Defects**
> A steel strip is produced in a complex process chain. At the end of the chain, the surface is checked with a camera. Errors are found, scratches, notches, cracks, and other structures that devalue the product.
> If you know the causal relationship between the error and the responsible process variables, you can specifically work to avoid the error. Often, identifying the process in which the actual error occurs is already interesting. ◄

Causality is the relationship between cause and effect. In technical applications, the causal mechanism is often given by physical processes. In the above example, for instance, a mechanical defect on a roller could be the possible cause of the surface problems. So, if you want to analyze causality, you need to take a closer look at the process chain.

Causal Classification

Fig. 7.8 helps us explain the causality of a process chain using an abstract example of anomaly detection. We consider a process chain with the processes P1, P2 to P6. Also, P3 is susceptible to a certain anomaly, which is later discovered in P6. The anomaly can therefore only be causally linked to the processes P1 and P2 or unknown effects that occurred before P1.

After the event has occurred, processes P4 and P5 can be used for its detection. For simplicity's sake, we assume that the event is a process anomaly and we are initially interested in detecting this anomaly. P6 is a specific test process, a quality check, for which we assume that it reliably determines whether the anomaly has occurred or not by using a special measurement. Note that P6 can therefore be considered an abstract label generator for supervised learning.

We can now identify various types of machine learning algorithms based on their causal context and define them by the abbreviation $\mathcal{A}_i$. The index i indicates the execution of the algorithm within a certain time segment. Algorithms of type $\mathcal{A}_4$ would, for example, have access to all data generated up to process P4. These causal relationships are important because if we only discover the anomalous event with $\mathcal{A}_5$ in P5, this algorithm would be almost useless. It only finds the anomaly shortly before P6, where it would be discovered in any case. In contrast, algorithms of type $\mathcal{A}_3$, which recognize the anomaly when it occurs, could help adjust processes P4 or P5 in time to save the product. The earlier the algorithms detect the anomaly, the better for the production path. Both $\mathcal{A}_4$ and $\mathcal{A}_3$ are detection algorithms, as they observe the anomaly after it has occurred.

Of course, it would be much better—if possible—to find algorithms of type $\mathcal{A}_2$ or even $\mathcal{A}_1$ that predict the occurrence of an anomaly before it happens. In Fig. 7.8, this is referred to as prediction. Here too, it is easy to see that the earlier algorithms of type $\mathcal{A}_1$ are more advantageous than $\mathcal{A}_2$, as their prediction occurs early enough to possibly change P2 and thus prevent the anomaly.

We can conclude this line of thought with the following statement: Every prediction or detection algorithm in the process industry should work with the smallest possible amount of data and at the earliest possible process step to ensure advantageous application.

- $\mathcal{A}_1$ can be trained with historical data from P1 and P5, where P6 provides the labels. Later, $\mathcal{A}_1$ is used to predict the occurrence or rather the probability of the anomaly, which contributes to the adjustment of P2.
- $\mathcal{A}_2$ can be trained with data from P1-P2 and labels from P6. It does not offer the potential to prevent the anomaly, but still represents an early detection.

As apparent as this seemingly simple relationship may seem, we often find projects and solution strategies for the use of machine learning in industry that are not aware of the causal chain.

7.3.4 Business Understanding

Here, decision-makers are the audience and they should neither get technical terminology nor overloaded explanations. Especially for the evaluation phase of the use of machine learning, this level is of importance.

For decision-makers, clearly measurable key performance indicators (KPIs) are important. This was already discussed in the CRISP-DM model. The KPIs quantify the success of a procedure. They can accompany the use of a procedure during commissioning to ensure continuous monitoring of the solution. In addition to the KPIs, the risk of a method is important. It answers the question of how sure one is that a procedure really has positive effects.

Summary
This final chapter brings our previous methods together to the overarching idea of explainable learning procedures. We use semantic technologies such as taxonomies and ontologies to capture digital descriptions in terms of data. An automated procedure can then access these descriptions.

In doing so, we have learned two classes in Python code that represent a taxonomy and a compact micro-ontology. They are intended to show you how you can conceptually implement your own solutions. Often, no own codes are used, but finished programs and languages (OWL) in which one models semantically.

An ontology does not only store descriptions. Rather, information on preprocessing or hyperparameters can also be stored here. A learning procedure can query the best processing chain via an ontology, depending on the type of input variables.

Explainability must ultimately always refer to a diverse audience. To implement modern machine learning in a technical environment, various levels in companies must be included. From the decision-maker in management to the process experts to the data experts—an algorithmic solution must be explainable for these different groups of people.

Tasks

7.1 Perform the sensitivity analysis in Listing 7.6 not only for the variables x[1] and x[2], but also for the variable x[0]. Is this dependency linear?

7.2 Perform the scan for input sensitivities also for the regression network in Sect. 4.5.8. Which variable has the strongest influence?

References

1. M. Babik and L. Hluchy, "Deep integration of python with web ontology language," in *2nd Workshop on Scripting for the Semantic Web*, 2006.
2. O. Bastani, C. Kim, and H. Bastani, "Interpreting blackbox models via model extraction," 2019.
3. U. Bhatt, A. Weller, and J. M. F. Moura, "Evaluating and aggregating feature-based model explanations," 2020.
4. U. Furbach and C. Schon, "Deontic logic for human reasoning," *Advances in Knowledge Representation, Logic Programming, and Abstract Augmentation*, pp. 63–80, 2014.
5. U. Furbach, C. Schon, and F. Stolzenburg, "Automated reasoning in deontic logic?" in *Proc. MIWAI 2014: Multi-Disciplinary International Workshop on Artificial Intelligence*, 2014.
6. D. Janzing, L. Minorics, and P. Bloebaum, "Feature relevance quantification in explainable ai: A causal problem," in *Proceedings of the Twenty Third International Conference on Artificial Intelligence and Statistics*, ser. Proceedings of Machine Learning Research, S. Chiappa and R. Calandra, Eds., vol. 108. PMLR, 26–28 Aug 2020, pp. 2907–2916. [Online]. Available: https://proceedings.mlr.press/v108/janzing20a.html.
7. M. A. Musen, "The protege project: A look back and a look forward," *Association of Computing Machinery Specific Interest Group in Artificial Intelligence*, vol. 1, no. 14, 2015.
8. P. F. Patel-Schneider, P. Hayes, and I. Horrocks, "OWL web ontology language: Semantics and abstract syntax," W3C, W3C Recommendation 10 February 2004, February 2004. [Online]. Available: http://www.w3.org/TR/owl-semantics/.
9. M. T. Ribeiro, S. Singh, and C. Guestrin, ""why should i trust you?": Explaining the predictions of any classifier," in *Proceedings of the 22nd ACM SIGKDD International Conference on Knowledge Discovery and Data Mining*, ser. KDD '16. New York, NY, USA: Association for Computing Machinery, 2016, pp. 1135–1144. [Online]. Available: https://doi.org/10.1145/2939672.2939778.
10. L. S. Shapley, *A Value for N-Person Games*. Santa Monica, CA: RAND Corporation, 1952.
11. S. Tan, G. Hooker, P. Koch, A. Gordo, and R. Caruana, "Considerations when learning additive explanations for black-box models," 2021.
12. E. Štrumbelj and I. Kononenko, "Explaining prediction models and individual predictions with feature contributions," *Knowl. Inf. Syst.*, vol. 41, no. 3, pp. 647–665, dec 2014. [Online]. Available: https://doi.org/10.1007/s10115-013-0679-x.

Appendix A: Basics in Python

A.1. Setting up a Python Environment

To be able to execute the codes presented in the book, you first need a Python environment on your computer. You have the free choice of several providers. We recommend installing the Anaconda Distribution[1], which is available for every operating system. It already includes most of the libraries we need for our codes.

Also included in Anaconda is the programming environment Jupyter[2]. This environment is well suited for getting started with machine learning. It is based on notebook files, which execute code fragments in the form of cells either in sequence or individually. You can modify the Python code in the notebook and execute the respective cell directly. Any error is quickly noticed and can be found quickly by individual cells.

A.2. Typing

The importance of the type of data has already been extensively discussed in the first chapter. In programming languages, the choice of the correct data type plays a prominent role. Many languages therefore require a clear definition of a variable to a data type. Such an agreement could be `Firstname=string`.

Python, on the other hand, works with dynamic typing. This means that the interpreter assigns the type of the variable while executing a program. The interpreter only learns the type from the first assignment: from `Firstname='Hans'` it concludes that you are using a `string` as a first name.

---

[1]https://www.anaconda.com

[2]https://jupyter.org/

M. J. Neuer, *Machine Learning for Engineers*, https://doi.org/10.1007/978-3-662-69995-9

Listing A.1 Different data types in Python

```
# integer
i = 1

# float
a = 2.0

# string
Name = 'Hans'
Color = 'Red'

# bool
isBlue = False
isRed = True

# Arrays
t = [0,1,2,3] # integer
x = [1.0,2.0,3.0,4.0] # float
persons = ['Hans', 'Peter', 'Frank'] # strings
```

In Listing A.1, lines 1–13 provide some examples of basic types. Lines 15–18, on the other hand, show vectors or `arrays`, which can contain multiple elements of the same type. Some programmers, especially those who work a lot with compiler languages, are bothered by the lack of type specifications. However, in later versions of Python, there is also a way to explicitly specify types, albeit only for documentation purposes. In Listing A.2, the same group of variables is created again, but now with explicit indication of the types in the code.

Listing A.2 Annotating types

```
# integer
i: int = 1

# float
a: float = 2.0

# string
Name: str = 'Hans'
Color: str = 'Red'

# bool
isBlue: bool = False
isRed: bool = True

# Arrays
t:list = [0,1,2,3] # integer
x:list = [1.0,2.0,3.0,4.0] # float
persons:list = ['Hans', 'Peter', 'Frank'] # strings
```

The type specification helps you especially when you later check the code automatically. There are tools that can better find consistency, type changes, and errors if the typing has been clearly defined.

In addition to the basic types `int`, `float`, `string` and `bool` as well as the `arrays`, Python also allows us to define heterogeneous structures. An important type for this is the `dictionary`. It represents a template for semi-structured data and allows complex data models to be written in a simple way. An example of this is shown in Listing A.3.

Listing A.3 Dictionaries

```
myDictionary = {'DictionaryID':'MyDict',
                'Color':'Red'
                'IsRed':True,
                'NumberOfPersons':3,
                'MeanAge':34.25,
                'Persons':['Hans','Peter', 'Frank']}
```

A.3. if-Condition, for-Loop and while-Loop

Like all programming languages, Python is capable of interpreting conditions. This is exemplified in Listing A.4.

Listing A.4 if-Condition

```
i = 10

if i == 10:
    print('i␣ist␣10')
elif i == 5:
    print('i␣ist␣5')
else:
    print('i␣ist␣alles,␣aber␣nicht␣10␣und␣nicht 5')
```

Listing A.5 shows a classic for-loop that runs from 0 to 9. It uses the internal command `range(start,end)`. Be careful, the output of this listing will end at 9. This is due to the definition of range(start, end), which starts at `start` and stops before `end`.

Listing A.5 for-Loop

```
for i in range(0,10):
    print(i)
```

for-loops iterate through their list. In the above case, this was a vector of numbers, in the next listing you will find a collection of names. In both cases, the for-loop goes through the vector element by element.

Listing A.6 for-each paradigm in Python

```
list = ['Hans', 'Peter', 'Karl', 'Thomas']
for eachName in list:
    print(eachName)
```

The while-loop is a variation of this concept.

Listing A.7 While-loop

```
i=0
while i<10:
    print(i)
    i+=1
```

A.4. List Abstraction/List Comprehension

List comprehensions are a powerful tool in Python for quickly generating new lists. They are often superior to for-loops. In Listing A.8 we find an example of a comprehension that creates a tuple (x, y).

Listing A.8 List comprehension

```
[(x,y) for x in range(0,2) for y in range(4,6)]
```

This notation is compact and allows lists to be written with little code. As can be seen from the example, they are clearly formulated, expressive, and easier to read. Despite all their advantages, for-loops are more didactically known. For algorithm prototypes in this book, we often choose the path via the classic loops. A significant reason is also the transfer to other languages, which is thereby simplified.

In some cases, we leave it to the exercises to transform code into the form of a comprehension.

A.5. Function Definition

Functions can be created using the note `def`. An example is in Listing A.9 the definition of a function that adds two numbers.

Listing A.9 Function definitions

```
def add(a,b):
    c = a+b
    return c

# oder kuerzer:
def add2(a,b):
    return a+b
```

Here too, the specification of types can help to make the code clearer. Listing A.10 therefore contains a variation of the function definition where explicit types are also specified.

Listing A.10 Annotated function

```
def add(a:float,b:float) -> float:
    c:float = a+b
    return c
```

A.6. Viewing Data

Now that we have dealt with the various ways in which we can label data within the programming language with descriptions, we want to show how to visually represent data. For us, it is of primary interest to get an overview of our data. In Python, the matplotlib library is a handy tool for plotting and visualizing. In Listing A.11, we see how the previous example 1.2 can be graphically output.

Listing A.11 Visualization of data with Matplotlib

```
plt.plot(myDataFrame['Time'],myDataFrame['Position'],'k-')
plt.scatter(loadedDataFrame['Time'],loadedDataFrame['Position'
    ], s=100, marker='o', color='b')

plt.xticks(fontsize=18)
plt.yticks(fontsize=18)
plt.xlabel('Time⎵/⎵.u.', fontsize=20)
plt.ylabel('Position⎵/⎵a.u.', fontsize=20)
```

Fig. A.1 shows the output of the diagram when the listing is executed. Matplotlib has a wide range of display types.

A.7. Class Definition

Python is an object-oriented language. You can define classes and based on this definition, generate multiple object instances that behave as prescribed in the class.

Fig. A.1 Output from the
program code in Listing A.11

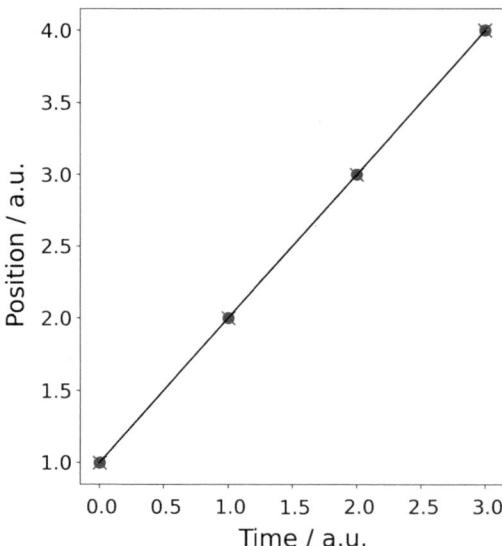

Listing A.12 Class definition

```
 1  # Klassendefinition
 2  class myClass():
 3
 4      def __init__(self, inputName): # Konstruktor
 5          self.name=inputName  # Klassenvariable
 6          i = 3        # lokale Variable nur innerhalb
                __init__
 7          self.i = 4 # Klassenvariable
 8
 9      def whoAreYou(self): # Klassenfunktion
10          return self.name
11
12  # Test der Klasse
13  myInstance = myClass(inputName='Torben')
14  myInstance.whoAreYou()
15
16  mySecondInstance = myClass(inputName='Hans')
17  mySecondInstance.whoAreYou()
```

In A.12, it is shown how to create a class and instantiate it. The term `self` plays
an important role. All variables that are defined via `self.?` are available throughout the entire class. Simple variables, as defined in line 6, are only available in the
section of the program where they were created.

Almost all algorithms that we will learn are created as objects.

A.8. Saving and Loading Data

Let's take a first look at ways to save and load data. This is indeed one of the fundamental first steps in any data mining work or developments for machine learning. Either we establish a connection to a database or we load data that is stored in special files and contains relevant data to our problem.

In Listing A.13, for simplicity's sake, we only assign a vector. We pick up an example from further above. The vectors represent the time t and a certain position vector x:

Listing A.13 Loading and saving data with Pandas

```
import pandas

t = [0,1,2,3] # index
x = [1.0,2.0,3.0,4.0] # measurement
myDictionary = {'Position':x, 'Time':t}

myDataFrame = pd.DataFrame(myDictionary)
myDataFrame.to_csv('Test.csv')
loadedDataFrame = pd.read_csv('Test.csv')
```

Listing A.13 shows on the one hand in lines 3–8 how you can save data and in line 9 how you can load the saved information again. Please note that in this example we store the data as a csv file, i.e., as comma-separated values.

An equally simple method to store entire dictionaries is provided by the Python function `pickle`. It stores the data in binary form and compresses it. This makes even large amounts of data compact and easily exchangeable. The A.14 shows the same example as A.13 and stores the data as a `pickle` file.

Listing A.14 Loading and saving Pickle files

```
import pickle

t = [0,1,2,3] # index
x = [1.0,2.0,3.0,4.0] # measurement
myDictionary = {'Position':x, 'Time':t}
pickle.dumps(myDictionary, open('Test.dat', 'wb'))

loadedDictionary = pickle.load(open('Test.dat', 'rb')
    )
```

The data for the examples in this book are provided in the `pickle` format. They can all be opened with the command in line 8 of A.14.